Optical Design for LED Solid-State Lighting

A guide

Series Editor

R Barry Johnson, a Senior Research Professor at Alabama A&M University, has been involved for over 50 years in lens design, optical systems design, electro-optical systems engineering, and photonics. He has been a faculty member at three academic institutions engaged in optics education and research, employed by a number of companies, and provided consulting services.

Dr Johnson is an IOP Fellow, SPIE Fellow and Life Member, OSA Fellow, and was the 1987 President of SPIE. He serves on the editorial board of *Infrared Physics & Technology* and *Advances in Optical Technologies*. Dr Johnson has been awarded many patents, has published numerous papers and several books and book chapters, and was awarded the 2012 OSA/SPIE Joseph W Goodman Book Writing Award for Lens Design Fundamentals (second edition). He is a perennial co-chair of the annual SPIE Current Developments in Lens Design and Optical Engineering Conference.

Foreword

Until the 1960s the field of optics was primarily concentrated in the classical areas of photography, cameras, binoculars, telescopes, spectrometers, colorimeters, radiometers, etc. In the late 1960s optics began to blossom with the advent of new types of infrared detectors, liquid crystal displays (LCDs), light emitting diodes (LEDs), charge coupled devices (CCDs), lasers, holography, fiber optics, new optical materials, advances in optical and mechanical fabrication, new optical design programs, and many more technologies. With the development of the LED, LCD, CCD, and other electro-optical devices, the term 'photonics' came into vogue in the 1980s to describe the science of using light in the development of new technologies and the performance of a myriad of applications. Today optics and photonics are truly pervasive throughout society and new technologies are continuing to emerge. The objective of this series is to provide students, researchers, and those who enjoy self-education with a wide-ranging collection of books that each focus on a relevant topic in the technologies and applications of optics and photonics. These books will provide knowledge to prepare the reader to be better able to participate in these exciting areas now and in the future. The title of this series is *Emerging Technologies in Optics and Photonics* where 'emerging' is taken to mean 'coming into existence', 'coming into maturity', and 'coming into prominence'. IOP Publishing and I hope that you find this series of significant value to you and your career.

Optical Design for LED Solid-State Lighting

A guide

Ching-Cherng Sun and Tsung-Xian Lee

Department of Optics and Photonics, National Central University, Taoyuan City, Taiwan

IOP Publishing, Bristol, UK

ISBN 978-0-7503-2368-0 (ebook)
ISBN 978-0-7503-2366-6 (print)
ISBN 978-0-7503-2369-7 (myPrint)
ISBN 978-0-7503-2367-3 (mobi)

DOI 10.1088/978-0-7503-2368-0

Version: 20220901

IOP ebooks

British Library Cataloguing-in-Publication Data: A catalogue record for this book is available from the British Library.

Published by IOP Publishing, wholly owned by The Institute of Physics, London

IOP Publishing, Temple Circus, Temple Way, Bristol, BS1 6HG, UK

US Office: IOP Publishing, Inc., 190 North Independence Mall West, Suite 601, Philadelphia, PA 19106, USA

Contents

Preface

LED solid-state lighting (SSL) has been intensively developed since the beginning of the 21st century. Among those developed technologies, optical design is one of the key issues in supporting SSL. To the details, SSL technologies can be divided into several divisions, such as materials, chip processing, packaging, optics, and human science. Optics technology, however, is a unique field that is not related to material engineering. This is somehow a problem for an optical designer working in SSL when reading a book related to SSL. There is a demand to offer a reader from the viewpoint of optics. This is the essential thinking in writing this book.

Working on optical engineering in SSL for more than fifteen years, Professor Sun and Professor Lee have accumulated abundant experience in handling various optical problems, which are important in pushing SSL toward completeness. The optical problems sometimes cover various ranges from material to human science. It is complex when one is preparing to teach a student and an engineer about optics technology in SSL. Therefore, we think that a book covering most optical issues for SSL is really necessary.

The writing work was started in 2018 and was completed in 2021. The book covers seven chapters. The first chapter introduces the background of SSL and knowledge for readers that four-level optical technologies will be discussed throughout the book. Besides, radiometry and photometry, the essential knowledge for handling illumination of SSL, is introduced to the readers. The second chapter briefly introduces basic optics to the readers. This is useful for a reader without an optics background, and it is also helpful to a reader in strengthening the necessary optical knowledge. The third chapter discusses the light extraction of an LED. This is an issue when the emission efficiency of an LED is concerned, and also is related to the light emission field owing to LED structures through the crystal epitaxy and chip processing. The fourth chapter discusses packaging technology, where phosphor modeling is one of the key issues. The reader will understand how to figure out the packaging efficiency when facing the related complex problem. The fifth chapter is important for an optical designer. The optical modeling technology of an LED light source is introduced with case studies. The sixth chapter discusses the secondary optical design, which is the main technique for an optical designer. The fundamental of energy transfer from a light source to a receiver is discussed. The seventh chapter introduces several different cases of optical design for LED SSL.

We believe that this book will be useful to a student learning optical technology, and helpful to an optical engineer in figuring out the optics of SSL.

This book was written by two authors. Without a doubt, it was hard work for the authors. Professor Lee needed to sacrifice his precious time accompanying his wife, daughter, and son while the two kids were in childhood. He would like to thank his wife, Ching-Yi, for taking care of the two kids in his absence. Professor Sun would like to thank his wife for her patience for his whole career working in optics, including writing this book.

We would like to thank all the students that drew the figures, found the references, discussed the problems, made the simulation, and read the manuscript. They are Chi-Shou Wu, Shih-Kang Lin, Quang-Khoi Nguyen, Thi-Thu-Ngoc Le in National Central University, and Yi-Ming Li, Yen-Lin Shih, Kai-Chun Chuang, Yang-Kuan Tseng in National Taiwan University of Science and Technology. Without their efforts, we don't think this book could be published in 2022.

This book is suitable for senior students in university or college or graduate students when encountering optical design or engineering for the first time.

Ching-Cherng Sun and Tsung-Xian Lee
10 January 2022
Taiwan

Author biographies

Ching-Cherng Sun

Ching-Cherng Sun received his BS in electro-physics from National Chiao Tung University in 1988 and his PhD in Optical Sciences, from National Central University (NCU), in January 1993. In 1996, he joined the faculty of NCU and became a Full Professor in 2002, named Distinguished Professor in 2006, and Chair Professor in 2014.

Professor Sun was presented the Industry Contribution Award for University by the Ministry of Economic Affair in 2009, the awards for Outstanding Technology to Industry, and the Outstanding Research Award by the National Science Council in 2009, the Outstanding Research Award by Taiwan Information Storage Association in 2012, the Invention Award for Photonics by Far Eastern Y Z Hsu Science and Technology Memorial Foundation in 2014, and the Outstanding Research Award by the Ministry of Science and Technology in 2015, and the Engineering Award of Taiwan Photonics Society (TPS) in 2018. He is a Fellow of the International Society of Optical Engineering (SPIE) and OPTICA (formerly OSA), the President of TPS (2019–2022), Professor Sun was one of the Chief Editors in *Journal of Solid State Lighting* (by Springer, 2014–2016), and was an Associate Editor of Applied Optics (by OSA, 2011–2017). He is currently an Associate Editor of OSA Continuum, Editorial member of Scientific Reports, and Crystals.

Professor Sun has authored and co-authored more than 170 refereed papers and 350 conference papers. He has been invited to give invited talks at more than 65 international conferences. His current major research includes holography, holographic storage, photorefractive devices, optical system, LED solid-state lighting, LED package, optical design, 3D scanning and modeling, and MR near-eye technology.

Tsung-Xian Lee

Tsung-Xian Lee, is the Associate Professor of the Graduate Institute of Color and Illumination Technology at National Taiwan University of Science and Technology. Professor Lee received his BS in physics at National Sun Yat-Sen University, and his MS and PhD in optics sciences at National Central University. His doctoral dissertation was on the topic of LED light extraction. He also worked for the EPISTAR Corporation to research advanced LED die and packaging technologies. He has worked for over ten years in the field of LED solid-state lighting. Presently, he is still committed to developing various LED applications, including intelligent lighting, human centrical lighting, healthy lighting, infrared sensing, and ultraviolet sterilization.

Chapter 1

Introduction of LED solid-state lighting

This chapter gives a brief overview of the background and the basis for the light-emitting diodes (LEDs). First, an introduction describes the origin of LEDs and their development. Next, the following sections provide you with the basics you need to get started, including how LEDs work, photon emission properties, and construction. In addition, this chapter also covers the basic concepts of radiometry and photometry, as this is essential knowledge when studying LED illumination optics. Finally, we explain the four-level optical design issues in LED lighting, which will be the main essence throughout the book.

1.1 The past, present, and future of LEDs

The development of human civilization is closely related to light. In the daytime, there is an irreplaceable natural light source from the Sun; at night, early people used fire, candles, oil lamps, and other heat sources to try to extend the time allowed for doing work and activity. Edison's commercialization of the incandescent lamp in the 19th century shed light on the dawn of the era of artificial electric light sources, and lighting gradually become necessary in people's life. With the development of lighting technology, many electric light sources have been developed and become popular, such as fluorescent lamps and discharge lamps. Although these light sources are now mature and have worked for mankind for more than 100 years, each has certain limitations. For example, incandescent lamps' luminous efficacy and lifespan, fluorescent lamps' mercury pollution, and discharge lamps' warm-up time are some of the often-criticized shortcomings.

At the end of the 20th century, the emergence of white LEDs led to a second lighting revolution, which is of unparalleled importance to mankind's history of lighting, just as important as when Edison invented the first commercial incandescent lamp. LEDs first originated in 1961 [1]. Dr Biard and Gary Pittman of Texas Instruments demonstrated an LED with gallium arsenide (GaAs) material, of which the emission wavelength range is near-infrared. In 1962, Nick Holonyak Jr of

doi:10.1088/978-0-7503-2368-0ch1

General Electric used gallium arsenide phosphide (GaAsP) to create the first practical application of visible LEDs with red light [2]. Since then, more colored LEDs have appeared. By 1991, HP and Toshiba developed a green LED made of aluminum gallium indium phosphide (AlGaInP) [3]. However, without blue LEDs, it was impossible to use LEDs of three primary colors of blue, green, and red for display and lighting applications. In 1993, Shuji Nakamura, who was working at Nichia, successfully developed high-brightness blue LEDs using gallium nitride (GaN) and indium gallium nitride (InGaN) [4]. This revolutionary invention broke through the technical bottleneck of the past 30 years and has turned a new page for the LED solid-state lighting (SSL). Soon after, a white light solution using blue LED and cerium-doped yttrium aluminum garnet (YAG:Ce) phosphor was introduced which fully promoting lighting technology innovation [5]. The 2014 Nobel Prize in Physics was awarded to three professors dedicated to blue LED research, namely Professor Isamu Akasaki from Meijo University, Professor Hiroshi Amano from Nagoya University, and Professor Shuji Nakamura from the University of California.

Compared with traditional incandescent and fluorescent lamps, LEDs have a longer lifespan, lower power consumption, higher switching speed, and lower driving voltage but without any heat radiation or mercury pollution, among others. These benefits are entirely in line with the current environmental protection trend. In addition, LEDs provide unique advantages of a compact form factor, vivid colors, and exceptional reliability, all of which make LED luminaires more diversified and novel [6–9]. Because of these, LEDs will undoubtedly be the dominant light source in the 21st century. In the future, all electric light sources may be entirely replaced by LEDs. However, at the beginning of the 21st century, white LEDs were still unable to meet the requirements of general lighting with respect to the luminous efficacy and light quality. Therefore, improving the brightness of LEDs became a hot topic, attracting many experts and companies worldwide to invest in the development of LEDs. In 2000, the luminous efficacy of white LEDs was only 30 lm W^{-1} [10], which was similar to that of halogen lamps. From 2001 to 2008, the LED efficiency has tripled and exceeded the 60–80 lm W^{-1} of fluorescent lamps. In 2016, white LEDs that reached 200 lm W^{-1} were launched [11], which is better than the other conventional light sources. However, a gap remains between market prices and consumer expectations when users not only have to be satisfied with the brightness, but also with the color appearance and lighting quality. How to improve the lighting quality and reduce the price has become a new challenge for LEDs.

Over the past ten years, many key technologies have achieved breakthroughs. White LEDs continue to deliver more lm W^{-1}, and their lm $\$^{-1}$ has also increased by nearly ten times from 2010 to 2020 [11]. Coupled with increasingly high electricity prices and environmental concern, consumers are willing to transition to energy-saving and environmentally-friendly LEDs as their primary light source. Nowadays, the application of LEDs is quite extensive, such as street lights, automotive headlights, stage lights, advertising signs, traffic lights, LCD backlights, flashlights, camera flashes, and decorative lights, to name but a few. You can see the application traces of LEDs inside. In addition, the value of LEDs goes beyond just replacing traditional light sources. Thanks to new applications such as intelligent lighting, health lighting, smart city, plant lighting, visible light communication, UV sterilization, and infrared sensing, LEDs have become more diversified and essential in our daily life. Moreover, the possibilities and advantages of LEDs allow us to re-examine well-known lighting requirements in the past and outline unknown lighting development for the future.

Continuous innovations and breakthroughs in LED technology will in turn promote the progress of other industries. The next stage of LED may enter the display field to replace LCD and OLED panels. A new breed of LED technology called 'MicroLED' is a way of shrinking the size of LEDs to pixel-level to create a color matrix display [12–14]. Compared with LCDs, MicroLEDs are self-luminous, which results in displays that provide better image quality (contrast, response time, for example) and higher power efficiency. In addition, MicroLEDs are made of inorganic materials. Compared with OLEDs, MicroLEDs are expected to be brighter, more durable, and potentially have a wider color gamut [15–17]. Besides, MicroLEDs will have a broader range of applications, including AR/VR/MR devices, head-up displays, and high-resolution wearable products, and many more. Therefore, MicroLEDs are regarded by the industry as the next-generation light sources for various applications, even though many difficulties are yet to be overcome in using MicroLEDs for commercial purposes.

The future is approaching, and in the end, it may be far beyond the future we have imagined. Looking at past experiences can provide us with what it takes to face the future challenges.

1.2 Working principle of LEDs

LED is a semiconductor device that can convert electrical energy into light energy. To understand the working principle of LEDs, we must first introduce the concept of semiconductors. Semiconductors are solid-state objects of which the electrical conductivity is between conductors and insulators. They are essentially a crystallized structure with atoms that are stacked up neatly. In general, when many atoms combine to form solid matter, there will be interactions between the peripheral electrons of the neighboring atoms. As the attractive and repulsive forces act simultaneously, a stable equilibrium state is finally reached. In the process of solid formation, some significant changes will be produced in the electron energy level, resulting in materials with different characteristics, such as conductors, insulators, and semiconductors. From the perspective of energy state, these stable electrons have lower energy and form a valence band. If some of the electrons obtain energy from the external actions, such as heating, illuminating, or electrifying, these excited electrons will then be free to move between atoms. Such free moving electrons can conduct electricity or heat. Because of their higher energy state, they form a conductive band. The energy difference between the conduction band and the valence band is called the 'band gap.'

Generally speaking, physical systems tend to go to a low-energy state, so, electrons first fill up the valence band. When electrons gain energy, chances are that they may transition from the valence band to the conductive band. Figure 1.1 illustrates the difference of band gap between insulators, semiconductors, and conductors. The electrons in the insulator are all in the valence band. Generally, the thermal energy of electrons is 0.025 eV at room temperature. In contrast, the energy band of insulators is more than 4 eV, so no electrons will appear in the conductive band. In the case of a conductor, the valence band and the conductive band overlap. In such a condition, electrons will form a 'sea of electrons' in the conductive band, efficiently conducting current.

The band gap of a semiconductor is smaller than that of an insulator (around 1–3 eV), and it is usually non-conductive at room temperature. Still, if the energy provided to it is higher than that of the band gap, the electrons inside have a chance to be excited to the conduction band. Once an electron jumps to the conduction band, it forms an electron vacancy called a hole in the valence band. Other valence electrons can fill that vacancy and produce another new vacancy. It would seem like a hole is moving in the valence band, just as an electron is moving in the conductive band. When an external electric field is applied, the negatively-charged electrons and the positively-charged holes will move in opposite directions to generate an electrical current.

On the other hand, the band gap concept makes it easy for us to understand the conductivity of semiconductors, and it can also be used to explain why semiconductors can emit light. Continuing from the above, those electrons that jump to the conduction band tend to return to the stable, low-energy valence band and fill the holes in a short period of time (about a few nanoseconds). This process is known as 'electron–hole pair recombination'. According to the law of energy and momentum conservation, since the electrons in the conduction band have relatively more energy, during the recombination process, the electrons must release additional energy through two main forms, including light or heat. LEDs generate light based on this semiconductor principle. The way LEDs emit light belongs to the realm of spontaneous emission. The phase, direction, and polarization of emitted photons are all randomly distributed. However, high-energy electrons interact with semiconductor atoms during the recombination process, such as changing the motion of the

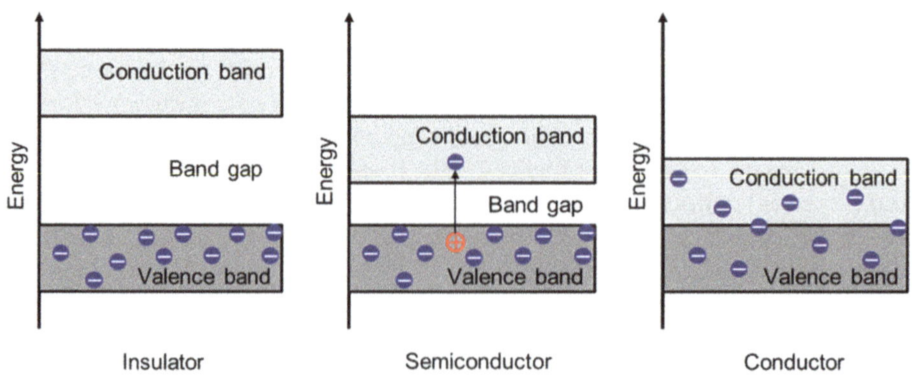

Figure 1.1. Difference between insulators, semiconductors, and conductors.

electrons or causing the atoms to vibrate. In that case, the way the energy is released often generates heat instead of emitting light effectively.

On the other hand, not all semiconductors can emit light. The type of the band gap of semiconductor materials includes 'direct' and 'indirect' ones. As mentioned above, a solid is composed of atoms stacked up in a certain way. This will cause electrons to move along different directions with different energies. As shown in figure 1.2, the energy of an electron will be a function of the electron's wavevector (k). Figure 1.2(a) is an energy band diagram with a direct band gap. The lowest energy of the conduction band and the highest energy of the valence band are at the same k point to allow electrons to fall directly from the conduction band to the valence band. In contrast, the energy band diagram of figure 1.2(b) shows that the lowest energy of the conduction band is located at the k point other than that one that corresponds to the highest energy of the valence band. Therefore, it will take some extra effort to make the electrons fall back to the valence band. This phenomenon tends to generate heat that makes the atoms vibrate instead of emitting light.

Which semiconductor materials are suitable for LEDs? According to element composition, commonly-used semiconductor materials can be divided into element semiconductors and compound semiconductors. On the periodic table of elements, group IV elements such as silicon (Si) and germanium (Ge) belong to element semiconductors. Each of their atoms has four periphery electrons. When they form a solid with the surrounding atoms, they can build the most stable eight-electron structure. However, because Si and Ge have an indirect band gap, they are not suitable for light-emitting devices. Another type of semiconductor is composed of more than one element, so they are called compound semiconductors, which can be further divided into III–V and II–VI compound semiconductors. III–V compound semiconductors are made with group III and group V elements. In group III elements, atoms are with three periphery electrons, such as aluminum (Al), gallium (Ga), indium (In), etc. In group V elements, atoms are with five periphery electrons,

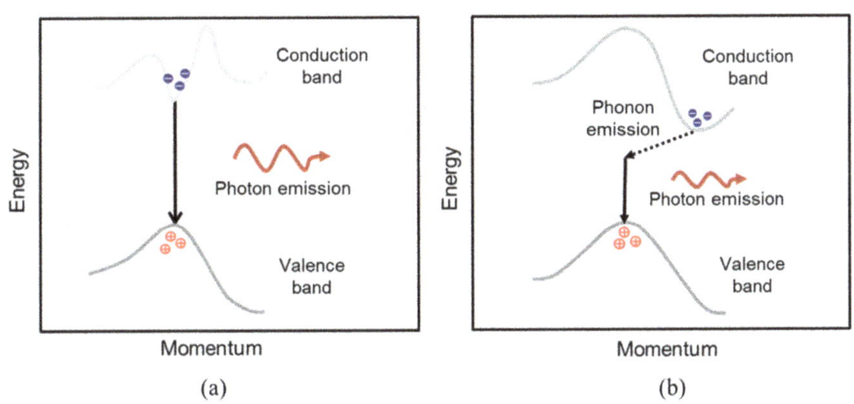

Figure 1.2. Simplified electronic band structure and optical transitions of semiconductors. (a) Direct-band gap semiconductor, (b) indirect-band gap semiconductor.

such as nitrogen (N), phosphorus (P), arsenic (As), etc. Some commonly-used ones are gallium arsenide (GaAs) and gallium nitride (GaN). The same goes with the II–VI compound semiconductors, which consist of group II elements, such as zinc (Zn), and group VI elements, such as oxygen (O) and sulfur (S). Some commonly-used ones are zinc oxide (ZnO) and zinc sulfide (ZnS), among others. III–V and II–VI compound semiconductors can also satisfy the requirement of the most stable eight-electron structure.

If compound semiconductors are classified based on the number of composition elements, they can be further divided into binary, ternary, quaternary and others. As the names suggest, binary compounds are composed of two elements, such as gallium arsenide (GaAs) and aluminum nitride (AlN); ternary compounds are composed of three elements, such as aluminum gallium arsenide (AlGaAs) and indium gallium nitride (InGaN); quaternary compounds are composed of four elements, such as aluminum gallium indium arsenide (AlGaInAs) and indium gallium arsenide phosphide (InGaAsP). The principle of these combinations is that the total sum of all atoms of group III must equal that of group V. Most compound semiconductors have a direct band gap, such as gallium arsenide (GaAs), gallium nitride (GaN), and indium phosphide (InP). There are also compounds of which the band gap is indirect, such as gallium phosphide (GaP) and aluminum arsenide (AlAs). Compound semiconductors are artificial semiconductors which grow compound thin film through epitaxial technology. The semiconductor materials required for LEDs must stack these compound thin films in sequence on a suitable substrate.

The band gap of semiconductor corresponds to photon energy, and therefore determines the wavelength of light. A larger band gap is required to emit light with a shorter wavelength (higher energy) and vice versa. The peak wavelength λ of light can be estimated by the semiconductor material band gap E_g in a unit of eV, i.e., $\lambda = 1240\ E_g^{-1}$, where the unit of λ is nm. If one wishes to generate visible light (λ range from 380 to 780 nm), then E_g should be between 3.26 and 1.63 eV.

Why do compounds with different elements and composition ratio change the band gap of semiconductors? Let us imagine when atoms of different sizes form a compound semiconductor, the distance between small atoms is shorter than that between large atoms. Generally, the smaller the distance between neighboring atoms, the more robust the bonding force will be, and it is not easy to separate. Therefore, the more small atoms there are in a compound semiconductor, the larger the band gap is. Conversely, compound semiconductors composed of large atoms have a smaller band gap. For example, GaN can emit blue light. If the smaller Al atoms in group III are used to replace the larger Ga atoms to form AlN, the band gap can be increased and can emit UV light. If a part of Ga atoms substitutes the larger In atoms to form InGaN, the band gap becomes smaller, and can emit green light. If all Ga atoms are replaced with In atoms to form InN, the band gap becomes smaller, and there is even a chance to emit IR light, as shown in figure 1.3.

So far, we have introduced the mechanism of light emission for semiconductors. Next, we will further explain the working principle of LEDs. The unique feature of semiconductors is that we can vary their conductivity behavior by controlling the

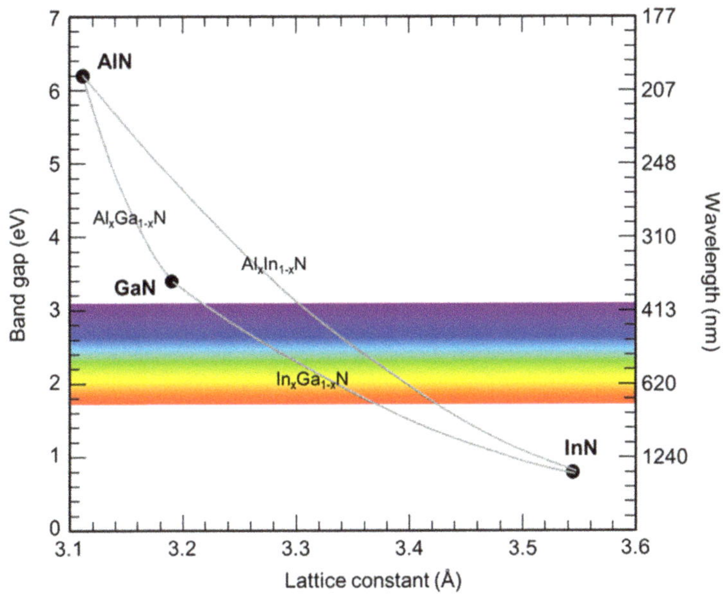

Figure 1.3. Direct band gaps in InN–GaN–AlN ternary alloy system as a function of lattice constant.

concentration and mobility of charge carriers (i.e., electrons and holes). As long as the intrinsic semiconductor is doped with a bit of impurity atom, its conductivity can be changed to achieve this purpose. Those containing excess positive charges (holes) are called P-type semiconductors, and those with extra negative charges (electrons) are called N-type semiconductors. When these two types of semiconductors are joined together, a PN junction is formed between them, and the holes will diffuse from the P-type to the N-type semiconductor. Over time, the electrons will also diffuse from the N-type to the P-type (diffusion: move from high to low concentration). When these mobile carriers leave, charged areas can arise, and then a built-in internal electric field will be formed at the junction to prevent the continued diffusion of electrons and holes. Only fixed charges remain at the junction in this equilibrium state that's called the depletion region. From the viewpoint of the band structure, due to the existence of the internal electric field, the conduction band and the valence band are bent, forming an energy barrier that prevents electrons from flowing from the N-type to the P-type at the conduction band, as shown in figure 1.4.

LEDs are essentially PN-junction diodes. As the name suggests, they have two electrodes and only allow current to flow in one direction. When we connect the positive side of the battery to the P-type semiconductor and the negative side to the N-type semiconductor, the bias given to the PN junction diode is called the forward bias, which allows the diode to conduct current. If we connect it in the reverse way, the bias given to the diode is called the reverse bias. In this case, the diode cannot conduct current. Once the reverse bias voltage is too high, it will cause the diode to break down, and the current rapidly increases in the reverse direction. Figure 1.5 shows the current–voltage characteristic of an ideal diode. For LEDs, we are

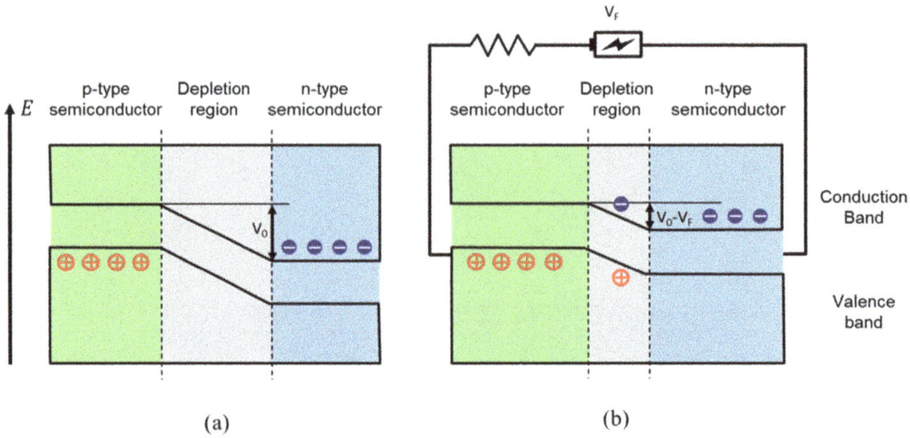

Figure 1.4. The formation of depletion region of PN junction diode and its energy band diagram (a) without forward bias, (b) with forward bias.

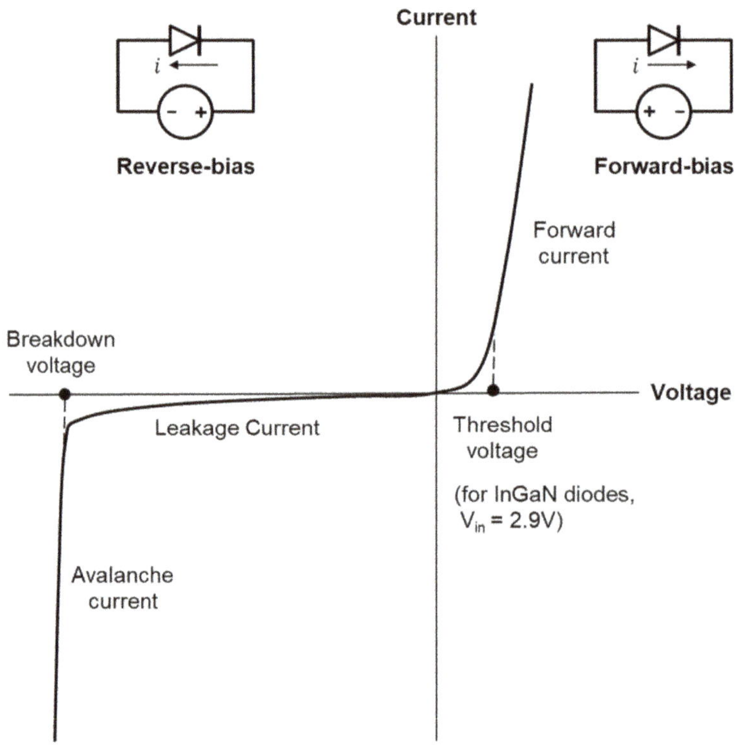

Figure 1.5. Current–voltage characteristic of diodes.

operating under the condition of forward bias. Electrons and holes are injected into N-type and P-type semiconductors, respectively. The electrons and holes will have a larger chance to recombine with others in a region around the PN junction, and the region for electron–hole recombination is called the active layer. The recombination process includes radiative recombination and non-radiative recombination. For radiative recombination, the electrons and holes will form the electron–hole pairs to release photons, and the energy of the photons is equal to the band gap of the semiconductor. For non-radiative recombination, however, the energy transforms to the vibration of the lattice. So, the energy converts into heat. Several physical mechanisms are involved when non-radiative recombination occurs, and material defects are the major mechanism. For high-efficiency LEDs, non-radiative recombination is undesirable. Thus, the material growth and chip processing are important to reduce the non-radiative recombination rate.

LEDs are semiconductor devices, and their light-emitting principle is to use the recombination of electron–hole pairs to generate light, unlike traditional tungsten light bulbs. The latter needs to heat the filament to a very high temperature (~3000 K) to radiate light. Therefore, LEDs will not consume a vast amount of energy but can still achieve the purpose of energy-saving. Moreover, the service life of LEDs is much longer than that of traditional light sources, so LEDs are a good choice in many places that are difficult to maintain. In addition, there are many advantages of LEDs that are beyond the reach of traditional light sources. For example, the LED size can be as small as a few millimeters to be designed with appropriate dimensions for different applications with a high degree of freedom of choice. Even though the spectrum of LEDs is narrower than that of most traditional light sources, the spectrum of an LED is broader than that of a laser, so LEDs are regarded as light sources emitting light in single color rather than single wavelength. The mixing of the three primary colors of light (red, green, and blue) can be used to create a light in various colors in a wider range to provide a wider color gamut of a display. Besides, LEDs have a fast response time, usually within microseconds, and can be used for optical communication.

However, if LEDs are to generate white light for lighting purposes, some work needs to be done for extending the colors. The practical white LED production method is based on the blue light emitted by GaN LEDs with yellow light converted by the phosphor to obtain white light. The mixture of blue and yellow light can be adjusted to perform white lights in different color temperatures [18, 19]. Alternatively, using blue LED to excite more phosphors of different colors such as green and red light, can further produce white LEDs with higher color rendering index (CRI) [20, 21]. Figure 1.6 shows the various ways for LEDs to make white light. The blue die with a yellow phosphor is the simplest and most effective way to generate white light. A blue die with multiple phosphors to have a more uniform spectrum is the way to generate white light with higher CRI. An ultraviolet (UV) die with blue/green/red phosphors is also for white light with higher CRI, but the UV light is always harmful to the packaging medium. That said, the reliability is an important issue and it could be neither a cost-effective nor high-efficient way for general lighting. Using blue/green/red LEDs to generate white light is not a common way in general lighting because the accurate injection current control is not easy and the CRI is not as good as for those with phosphors, though the mixed color is adjustable with this approach.

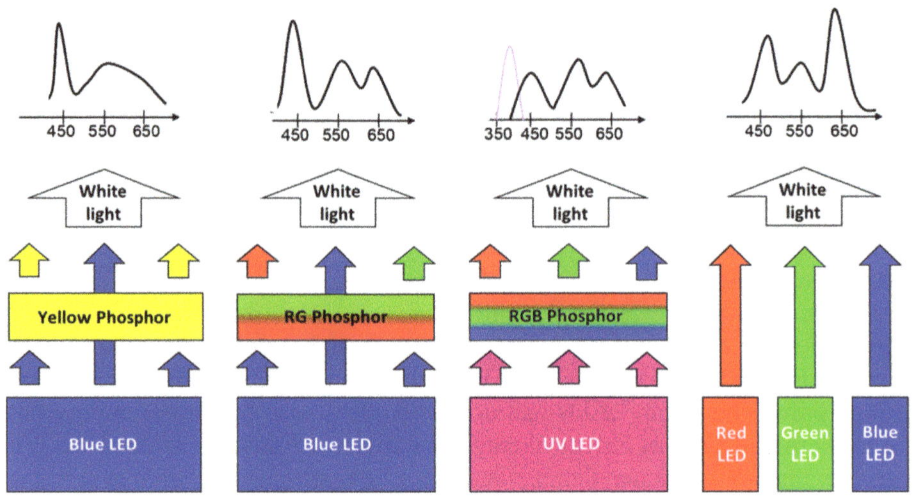

Figure 1.6. Four solutions to produce white LEDs.

Phosphors are solid-state luminescent materials. The principle of light emission is similar to the electroluminescence (EL) of LEDs. However, what the phosphor absorbs is not electric energy but the energy from electromagnetic radiation, and this is referred to as photoluminescence (PL). The phosphor photoluminescence can be classified into two types: fluorescence or phosphorescence. The distinction between the two depends on the phosphor's half-life. Generally, the half-life of fluorescence is about 10 ns–1 ms, and phosphorescence is about 1 ms–10 s, which is much slower than fluorescence. Typical phosphors used for LED lighting are inorganic fluorescent materials.

The above has introduced the general understanding of how LEDs make light. But this is just a starting point. The light generated from the PN junction must pass through a series of barriers to escape the LED before it is seen by a user. In the next section, we will discuss the barriers affecting the overall efficiency of LED lighting.

1.3 Luminous efficiency evaluation of LEDs

The overall luminous efficiency of LEDs is constrained by carrier injection efficiency, internal quantum efficiency, light extraction efficiency, packaging efficiency, and phosphor conversion efficiency. Improvements on each of these efficiencies are critical.

For an LED to emit light, the prerequisite is to excite the electrons in the semiconductor from the low-energy level to the high-energy level. The excitation of electrons relies on the injection of external carriers. The definition of carrier injection efficiency (IE) is the ratio of the number of carriers injected from the electrodes to the number of carriers obtained in the active layer. During the entire process between the carriers entry and departure of the semiconductor, the additional resistance of the LED device causes a higher operating voltage and consumes input electric power. Take GaN LED as an example. The additional resistance mainly comes from the contact electrode on the junction between the metal and semiconductor. They have to be small enough to reduce the loss of carrier injection. Therefore, increasing the doping concentration of p-GaN or using a contact electrode with a high work function can effectively improve the IE [22, 23].

With the injection of carriers, electrons and holes recombine in the active layer and release energy in the form of photons. In an ideal situation, the direct recombination of an electron–hole pair can radiate a photon. However, due to material defects and other factors, electrons and holes may not be able to recombine effectively. This would reduce the number of photons being radiated. The ratio between the number of photons and the electron–hole pairs is called 'internal quantum efficiency' (IQE). The IQE is primarily related to the material quality, epitaxial structure and heat dissipation mechanism. To maintain a high IQE, one of the essential conditions is to look for a substrate that can lattice-match with the epitaxial layer and reduce the material stress and defects. At present, GaAs-based LEDs apply the GaAs substrate with lattice-match, and its IQE is estimated to have reached nearly 100% [24–26]. However, for GaN-based LEDs, the lattice mismatch between GaN and the now-commonly used sapphire substrate is about 16.3%. Thus, there will be relatively high epitaxial defects. But through the improvements in epitaxial procedure, the IQE of current blue LEDs can reach more than 90% [27]. Despite the outstanding performance of epitaxial technology, there is still a gap between the overall luminous efficiency of LEDs and their IQE. The key lies in the impact caused by LED light extraction.

Light radiated from the LED die inside has to travel through many media and interfaces before reaching the external. Moreover, even if the light reaches the boundary, it may not be directly extracted. The ratio of the number of photons that can be extracted to the number of photons generated by the active layer is called 'light extraction efficiency' (LEE). The reason for limiting LEE mainly comes from the high refractive index of the LED die itself. For example, the refractive index of GaAs that produces red light is as high as 3.6, and GaN for blue/green light has a refractive index of 2.4. Both have a higher refractive index than that of general transparent media. Therefore, when the light reaches the LED die boundary, it will face the limitation of total internal reflection. The LED die is like a prison of photons, meaning only a tiny portion of the light less than the critical angle can escape. Fresnel reflection is another unavoidable optical loss. Even if the photon may escape successfully, some of its energy will be reduced. As a result, most of the photons trapped inside the LED die will eventually be absorbed by the LED itself, converting the energy of radiated light into waste heat. This heat will further reduce IQE, resulting in wavelength shift and shortened device life. Therefore, increasing LEE will help minimize heat generation. The improvement of LEE is mainly related to its LED die material, geometry, substrate, electrode, and packaging [28, 29].

With what is said above, the external quantum efficiency (EQE) is defined as the ratio of the number of photons emitted by the LED device to the number of injected electrons. It can be regarded as the combined result of injection efficiency, IQE, and LEE, as shown in formula (1.1):

$$\text{EQE} = \frac{\text{extracted pohton number}}{\text{injected electron number}} = \frac{E/h\upsilon}{I/e} = \text{IE} \cdot \text{IQE} \cdot \text{LEE}, \tag{1.1}$$

where E is the total output light energy; $h\nu$ is the energy of a photon; I is the input current; e is the charge of an electron. The luminous efficiency of LEDs can also be directly described by electro-optical power conversion called 'wall-plug efficiency' (WPE). It is defined as the ratio between the optical output power and the electrical input power. Since WPE can be measured directly, it is usually used as an evaluation index of LED luminous efficiency. The relationship between WPE and EQE is shown in formula (1.2):

$$\text{WPE} = \frac{P_{\text{out}}}{P_{\text{in}}} = \frac{P_{\text{out}}}{IV} = \text{EQE} \times \frac{h\nu}{eV}, \tag{1.2}$$

where P_{out} is the optical output power, P_{in} is the electric input power, I is the input current, and V is the relative voltage.

In addition, the visual perception of the human eye should be considered, especially for white light. The brightness response of the human eye is not linear, and it varies with different wavelengths of light. Therefore, the radiant flux must be converted into an effective quantity for the human eye using a luminosity function as the weight. For any given light source, its spectral power distribution (SPD) is $F(\lambda)$, then its luminous efficacy of radiation (LER) can be derived by using formula (1.3):

$$\text{LER} = \text{WPE} \times \frac{\kappa \int F(\lambda)V(\lambda)d\lambda}{\int F(\lambda)d\lambda}, \tag{1.3}$$

where the unit of LER is lumen per watt (lm W^{-1}), $V(\lambda)$ is the photopic luminous efficiency function, κ is a constant with a value of 683, which is the maximum luminance efficacy value for the monochromatic light at a wavelength of 555 nm. For the details, please refer to sections 1.1–1.5.

1.4 LED die and its package

A standard LED device is to seal the bare LED die inside a package body. The LED die is like tungsten in an incandescent bulb, which is the luminous core. The LED package function provides the necessary support for the electrical, thermal, and optical performances of the LED die. Some aspects include electrical connection, die protection, heat dissipation, and light extraction.

Therefore, besides the existing electrical and thermal requirements in other semiconductor technologies, the LED device design must also consider more critical optical matters. Some of the crucial points in the LED optical design are how to extract light more efficiently and distribute light more effectively, among others. Compared with the single-color LED package, the white LED packaging technology with mixed phosphors is more complicated. In addition to the problems mentioned above, the quality of white light must also be considered, such as color temperature, color rendering, and color uniformity. For the LED to meet the above requirements, the design of the die must work with the package, and both are indispensable. Nowadays, there are many types of LEDs that have been developed, but we will not examine them one by one here. This section will introduce the GaN LED die and its packaging, including the materials, structures, and parameters related to optics.

GaN LEDs can categorize the dies based on different substrates. The main functionality of the substrate is to carry high-quality epitaxial growth. GaN LEDs are different from other III–V LEDs. Due to the lack of a lattice-matched substrate for growing GaN, the epitaxial layer will have many defects which will affect its luminous efficiency and lifetime. Currently, GaN LEDs are produced with the sapphire substrate which has many advantages. First, the sapphire substrate has a mature manufacturing process and technology with stable quality of growth. Second, sapphire is perfectly durable and can be utilized in a high-temperature growth process. Third, sapphire has high mechanical strength and is easy to operate and clean. However, certain limitations lie with the use of sapphire as a substrate. Besides the lattice mismatch mentioned previously, sapphire is an insulator with a resistivity greater than 10^{11} Ω·cm under the normal temperature. It cannot be made into an LED with vertical electrodes. As a result, the P-type and N-type electrodes have to be created on the same side of the epitaxial layer, which means that part of the active layer must be removed from the die and this reduces the effective light output area, as shown in figure 1.7. Moreover, as the currents flow laterally, it is easy to cause current crowding and increase series resistance with concerns that the temperature of the components may rise. Also, sapphire does not have good thermal

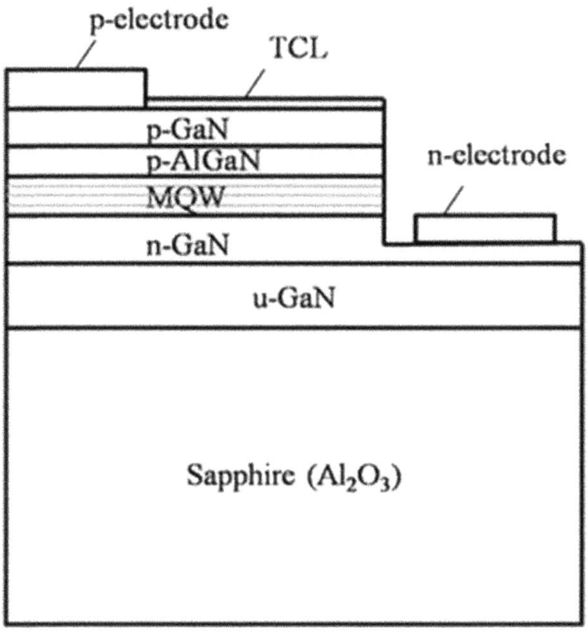

Figure 1.7. The GaN LED die structure with a sapphire substrate.

conductivity. With a thermal conductivity coefficient of 35 W $m^{-1}{\cdot}K^{-1}$ only, a large amount of heat cannot escape from the substrate easily, and the operating current of the LED is restricted. Thus, thermal conductivity is a reasonably important consideration for high-power LEDs. Nevertheless, sapphire remains the primary substrate material for GaN LEDs to balance the production cost and epitaxy quality.

Figure 1.7 shows the die structure for a typical GaN LED based on a sapphire substrate. The main manufacturing processes include epitaxy, etching, deposition, grinding, dicing and others. For epitaxy, metal-organic chemical vapor deposition (MOCVD) growth is usually used. In order to grow a high-quality GaN on a sapphire substrate, a thin layer of AlN or GaN has to be grown first for lattice matching under a low temperature. What is grown subsequently would be the n-GaN doped with Si, followed by a multiple-quantum well (MQW) of 5–15 cycles of InGaN/GaN as the active layer. A layer of AlGaN doped with Mg is grown as the cladding layer, followed by a p-GaN layer doped with Mg in the end. Moreover, a thin layer of ohmic contact electrode must be deposited on p-GaN to spread the current and make light generation uniform. This layer is called the 'transparent conductive layer' (TCL). Currently, indium tin oxide (ITO) is used as the TCL, and its light transmittance is around 90%, which can significantly improve the LED's brightness.

Once the epitaxial growth is completed, the die processing begins. For LEDs that are based on a sapphire substrate, processes such as photolithography and dry etching should first be used to etch the region needed for the *n*-type electrode. But for electrodes with a vertical structure, this step can be skipped. Then, a TCL, a *p*-type electrode, and an *n*-type electrode are sequentially fabricated. Finally, grind the substrate to reduce it to a thickness of 100 um to make it easier to dice into dies. The most common size of LED dies is between 0.2 mm and 1 mm.

Another substrate option for GaN LEDs is silicon carbide (SiC). The structure of its LED die is shown in figure 1.8. Many outstanding advantages are associated with the use of a SiC substrate. First, SiC has good conductivity and can be made into a vertical electrode LED without using a horizontal electrode. It not only increases the light generation area, but also improves the current uniformity. Second, SiC also has good thermal conductivity, and its thermal conductivity coefficient is 490 W $m^{-1}{\cdot}K^{-1}$, which is ten times higher than that of the sapphire substrate. This allows the heat generated inside the LED to dissipate effectively, making it more conducive to high-power LEDs. Third, the hardness of SiC is not high, so the LEE can be improved by die shaping. However, this increases the manufacturing cost as the current price of SiC is still too high.

Sapphire has poor electrical and thermal conductivity. These two shortcomings have a more noticeable impact on the performance of high-power LEDs. On the other hand, a challenge lies in using SiC substrates as its price is high. To overcome these difficulties, the substrate replacement process allows using materials with good electrical and thermal conductivity, such as Si and copper (Cu) and replacing sapphire substrates after epitaxial growth, as shown in figure 1.9. This type of LED has an opaque substrate, and the light escaping faces are reduced to five from six.

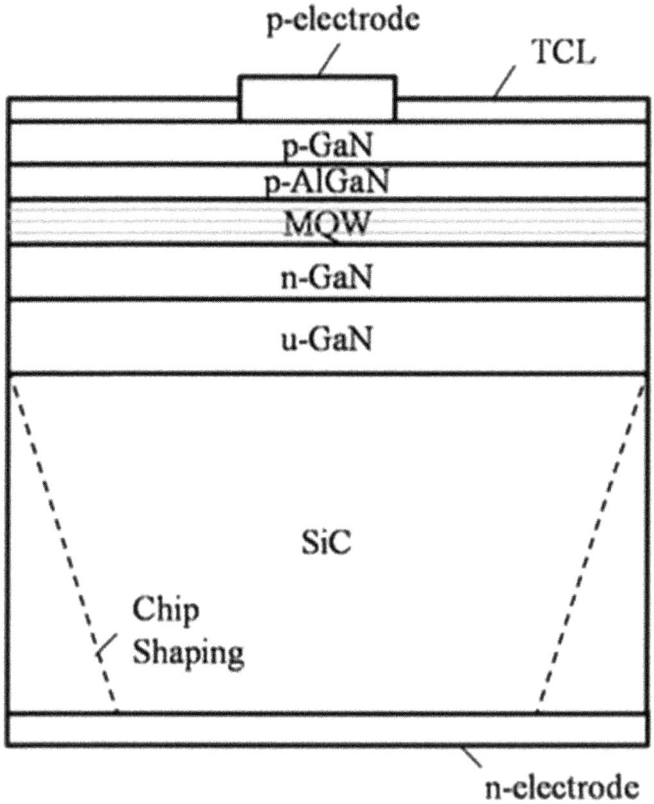

Figure 1.8. The GaN LED die structure with a SiC substrate.

Because the original substrate is removed, only a thin GaN epitaxial layer exists with the new substrate, so it is called ThinGaN structure [30]. The manufacturing method mainly uses wafer bonding and laser lift-off to replace the substrate. Besides, a highly reflective bottom mirror has been used to recycle the backward light and surface roughness/texture has been used to provide more changes of light escaping so that the LEE can be effectively increased. Chapter 3 will conduct a detailed analysis of the LEE of the above three LED die structures.

The LED packaging is developed based on the integrated circuit (IC) packaging technology. As the purpose is not the same, the packaging structure will also be different. In a general situation, the IC die is sealed in a package, of which the primary purpose is to protect the die, complete the circuit interconnection, and dissipate heat. However, the most significant difference in LED packaging is that the optical factor has to be considered. Besides the existing electrical and thermal parameters, the LED packaging also has to consider the optical issue. For this reason, it is not possible to apply the IC packaging technology entirely in LED packaging.

The LED packaging processes mainly include die bonding, wire bonding, and encapsulation. Depending on different application scenarios, shapes, dimensions,

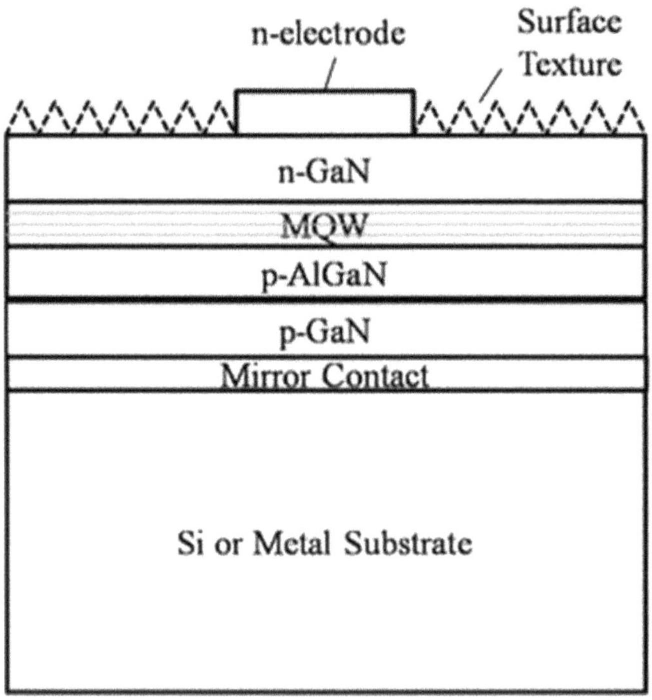

Figure 1.9. The ThinGaN structure.

heat dissipation schemes, and light output effects, there will be many different design types for LED packaging.

The earliest LED has a through-hole packaging. Its structure has two silver-plated lead frames as the main body. One has a reflector cup for LED die mount and phosphor filling, and the other is electrically connected by wire bonding, and the two lead frames are covered via epoxy molding. Its geometric shape is bullet-like, which can enhance the directivity of light. Since the main heat conduction path of the package is the lead frames, it is limited by its contact area. The encapsulation material is epoxy, and its resistance to heat and ultraviolet light is poor. Therefore, such a package structure is more commonly used for low-power LEDs such as indicator lights, as shown in figure 1.10.

With the demand for miniaturization of the electronic products such as computer, communication and consumer devices, a surface mount device (SMD) type of package has been developed. The feature of this package is that after the copper lead frame is stamped and electroplated, the lead frame is covered with injection-molded plastic to form a space for LED die mounting, encapsulating, and phosphor filling. This package is suitable for mid-power LEDs, and is widely used in backlighting and indoor lighting.

To obtain high brightness needed for general lighting, the development of LED dies tends to move towards bigger sizes to load higher current injection to increase the light output power. However, to maintain the stable luminous efficiency of the LED, the heat generated by the active layer must be effectively dissipated. In most

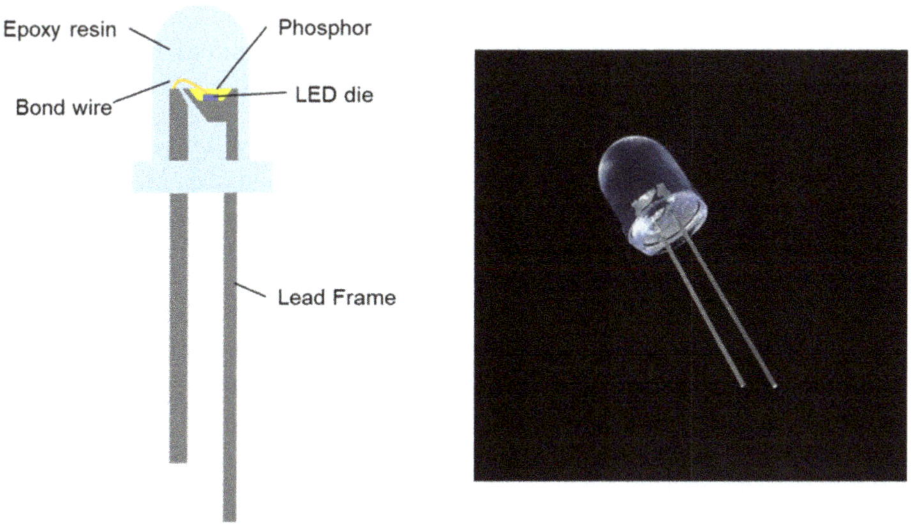

Figure 1.10. Low-power through-hole LED package.

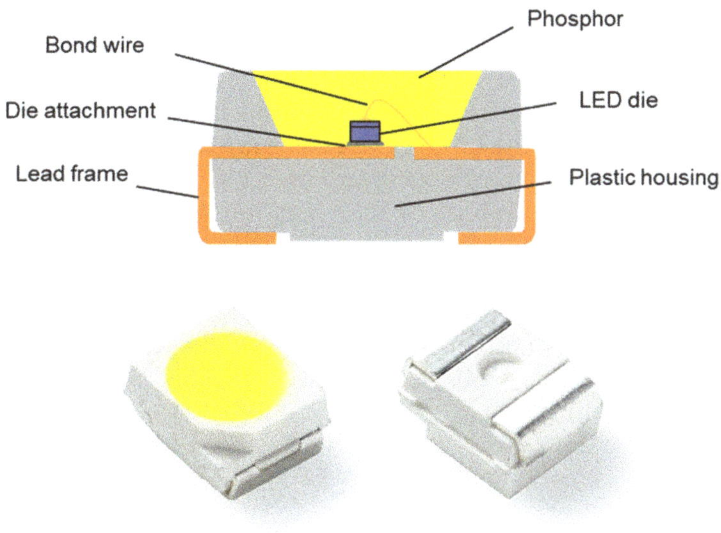

Figure 1.11. A mid-power SMD LED.

cases, the thermal conductivity of the LED package is poor, so one will worry about the reliability of the materials inside the package, such as broken gold wires or yellowing of glue. Furthermore, the luminous efficiency will drop significantly as the temperature rises, resulting in shortened device life or thermal drift in wavelength and voltage, as shown in figure 1.11.

The emergence of high-power LEDs has pushed up the penetration rate of LEDs in lighting. Whether it is the LUXEON K2 series launched by Lumileds, or the later

Cree XP series and Lumildes REBEL series, the common feature is the vastly reduced thermal resistance of the package [31, 32]. The initial high-power LED package is to place a die reversely in the sub-mount with electrostatic discharge protection and then to mount on a heatsink slug based on either aluminum or copper. Such a package avoids the poor thermal and electrical conductivity of the sapphire substrate and uses the high thermal-conductive metal slug to further reduce thermal resistance. Also, this so-called flip-chip packaging can prevent the electrodes from blocking the path of light extraction and increase the luminous efficiency. With respect to the optics, a transparent hemispherical silicone lens is used to prevent the gold wire from breaking due to thermal expansion and contraction. The shape of this hemispherical lens is also quite helpful for LED light extraction. With the advancement of technology and materials, the thermal resistance of high-power LEDs continues to reduce, and the input power is much higher than before. For example, ceramic substrates can further solve the risk of potential deterioration of material adhesion, thermal expansion coefficient mismatch, and poor long-term reliability. In addition, methods such as ThinGaN and thin-film flip-chip are much better than flip-chip in thermal conductivity, as shown in figure 1.12.

Another package type that can provide higher power is called chip-on-board (COB). The package can supply the maximum light output power per unit emitting area by increasing the number and density of LED dies. Generally, the metal core PCB (MCPCB) is used as the substrate for mounting dies covered with phosphor. A COB LED has a simple package structure, and it can even be integrated with an IC circuit. In addition, the COB LED can easily achieve high lumens per area (lm A^{-1}). From the viewpoint of optical design, it has many advantages. However, the heat density could be too high to keep high luminous efficiency, making heat dissipation a potential issue to the users, as shown in figure 1.13.

To omit the current packaging processes of die bonding and wiring, flip-chip LED incorporating wafer-level packaging technology has been proposed, and it only requires an LED die and phosphor. This new-generation device is called 'chip-scale package' (CSP) LED [33, 34]. Compared with the other LED packaging approaches, this novel method has a relatively simple process and can effectively reduce the packaging size and cost. In addition, a CSP LED has the advantage of a

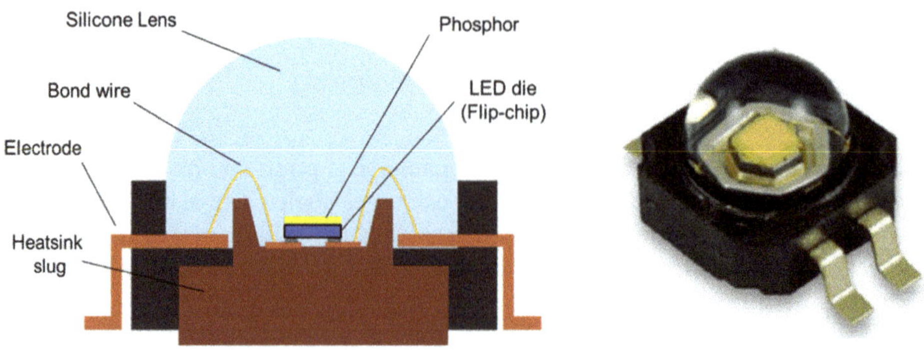

Figure 1.12. Watt-level high-power LED package (Lumileds' LUXEON K2).

Figure 1.13. High-brightness COB LED.

Figure 1.14. Miniaturized CSP LED package.

wider emitting angle attributed to their transparent sidewalls. Besides, CSP LEDs are always slim, and this feature makes them suitable for backlight products [35], as shown in figure 1.14.

1.5 Radiometry and photometry

Radiometry is a way to judge or evaluate how much electromagnetic wave is emitted by a light source or is incident on a receiver. When the electromagnetic wave falls into the visible range, the detection quantity needs to fit the human-eye response, and to evaluate the brightness by human eyes, so radiometry is transferred to photometry. The natural radiation into the Earth is sunlight, which can be regarded as the radiation of black body. Black body radiation is the radiation of electromagnetic waves when the radiator is an ideal black body, where the radiation emissivity is 100%. Thus the radiation spectrum can correspond to the surface temperature of the black body. Figure 1.15 shows the black body radiation at different temperatures of the light source [36]. The peak wavelength of the radiation spectrum moves to a longer wavelength when the temperature of the radiator is lower. Wien's law cleverly describes the phenomena

$$\lambda_{\max} T \approx 3000\ \mu\text{m K}, \tag{1.4}$$

where $\lambda_{\max}$ is the peak wavelength in μm, T is the temperature of the radiator, K is the absolute temperature.

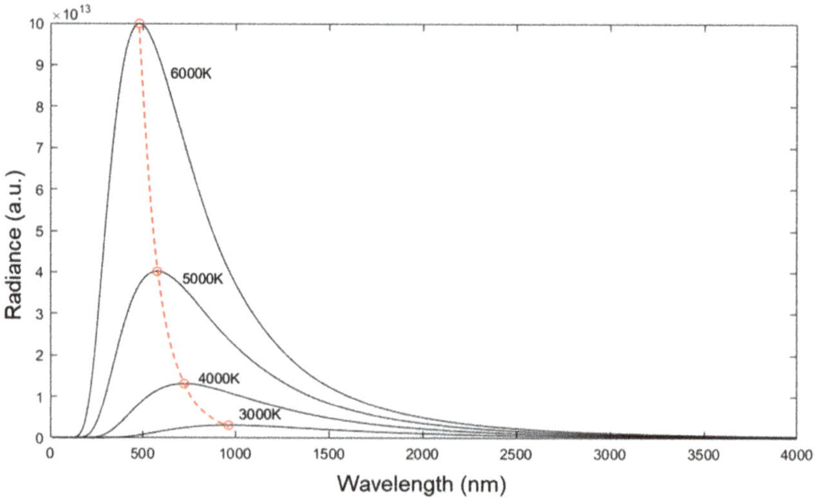

Figure 1.15. Wien's law addresses that the maximum wavelength of radiance of a black body times the temperature is around constant.

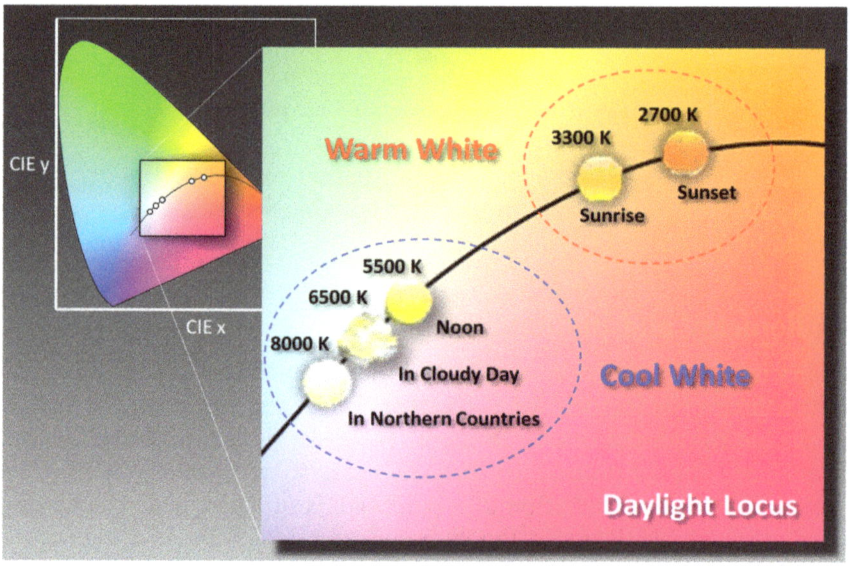

Figure 1.16. The black body radiation curve in the chromaticity coordinates.

Color of sunlight changes from morning to evening, but is always regarded as white light. Therefore, the definition of white light can follow the black body radiation curve in chromaticity coordinates, where the temperature of the black body is in the range of 2500 K to 10 000 K, as shown in figure 1.16. Since the color sensation of the white light corresponds to the temperature of the radiator, we may describe the specific white light along the black body radiation with the corresponding temperature, so it is called correlated color temperature, and simplified as CCT [36].

In order to check the radiation spectral distribution, we have to know the wavefront characteristic of an electromagnetic wave. The ideal wavefront can only be generated by an ideal point source, and the wavefront is a perfect sphere. As the observation distance becomes larger, the sphere will be enlarged. If just a part of the sphere is inspected, the partial sphere is a surface with a radius of curvature equal to the propagation distance. Here the partial surface is called a spherical wavefront. An ideal spherical wavefront can be collimated with a positive lens when the convergent lens can bend the spherical wave to a planar wave. The spherical wavefront and the planar wavefront are both important in radiometry.

As shown in figure 1.17, a point source radiates a spherical wave. Confining in a solid angle (Ω), which is defined as the ratio between the area of the cross section and the square of the distance [37], the optical flux does not change no matter how far the wavefront is propagated. Such a quantity is called intensity, which is written

$$I = \frac{dF}{d\Omega}, \tag{1.5}$$

where F is the flux confined in the solid angle.

For an ideal point source, the intensity is constant and independent of the propagation distance and the propagation direction. To evaluate the received radiation, another quantity called irradiance (E) is defined [37], and can be written

$$E_0 = \frac{dF}{dA} = \frac{I}{R^2}, \tag{1.6}$$

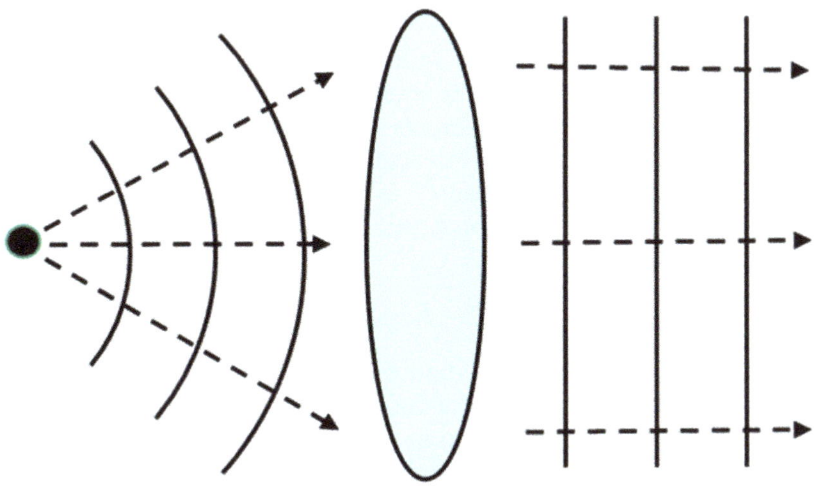

Figure 1.17. A planar wavefront can be transferred from a spherical wavefront with a positive lens.

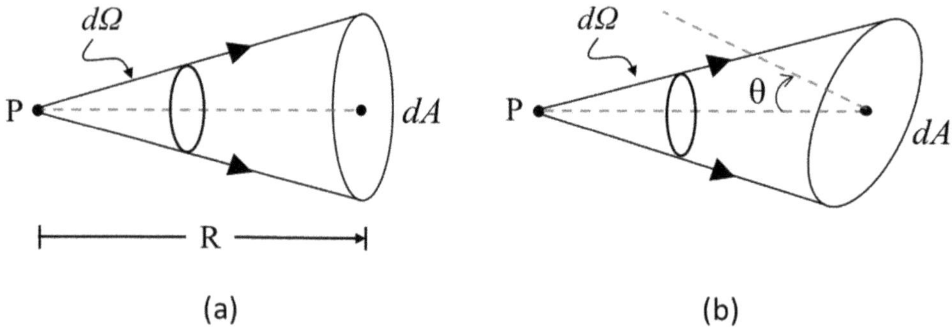

Figure 1.18. (a) The light cone with a cross section of dA, and (b) of $dA \cos \theta$.

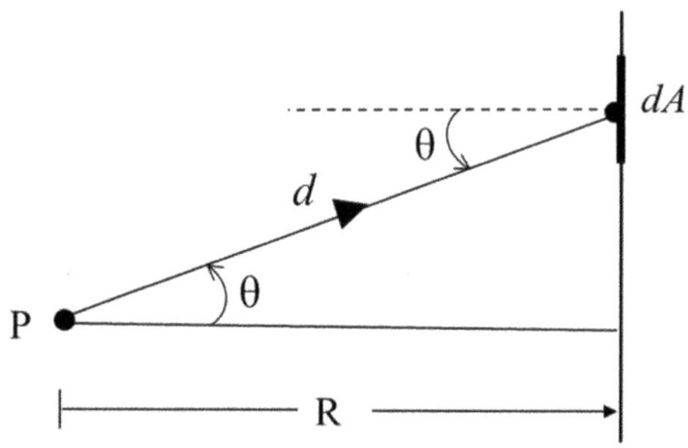

Figure 1.19. The normal direction of the irradiated surface does not point to the light source.

where A is the cross section of the radiated surface and R is the propagation distance. Since A is proportional to the square of propagation distance, E is inverse square proportional to the propagation distance. Such a property is dominant for a point source or a light source that looks small. If the cross section makes an angle θ with the optical axis of the light source as shown in figure 1.18, the area of the slanted cross section is inversely proportional to a factor of cos θ, so that the irradiance becomes

$$E(\theta) = E_0 \cos \theta. \tag{1.7}$$

This is called cosine first law for irradiance [37].

The measurement of the irradiance for a point source must be careful. If the normal direction of the effective detection surface of a detector does not point to the light source as shown in figure 1.19, the irradiance becomes

$$E_\theta = E_0 \cos^3 \theta. \tag{1.8}$$

If a light source has an un-neglected area, it should be regarded as an extended light source. An ideal extended light source consists of many ideal point sources. The intensity is no longer constant along all angles owing to the un-neglected light source area. The measurement quantity will vary upon different viewing angles. The effective area of the extended light source along a larger viewing angle becomes smaller, and then the measured intensity will become lower. Thus a new measurement quantity radiance is defined

$$B = \frac{dI}{dS\cos\theta}. \tag{1.9}$$

Since the factor of slanted area, $dS\cos\theta$, is normalized in the definition of radiance B illustrated in equation (1.9), radiance is constant for an ideal extended light source, which is called Lambertian light source, as shown in figure 1.20. A Lambertian light source can be regarded as the ideal extended light source and is an important reference when a specific surface of the radiator is evaluated. In drawing a radar figure to see the intensity as a function of viewing angle, the angular intensity distribution of a Lambertian light source is a circle. The half width is found ±60°. Therefore, a Lambertian light source will perform an intensity pattern with FWHM (full width at half maximum) angle of 120°, as shown in figure 1.21.

To judge if a surface is a Lambertian one, a human eye is one of the best tools. A human eye is an imaging system. The irradiance on the retina corresponds to a feeling of brightness radiated/reflected by an object. A Lambertian light source seen by a human eye will have equal brightness no matter how far the extended light source is or how large the viewing angle is. Thus, for a Lambertian reflective surface, one cannot judge where the illumination light source is incident on the surface. If the surface can be seen as a specular reflection from a certain viewing angle, the surface is not a Lambertian surface. When the light source is a Lambertian one, the measurement of irradiance is very sensitive to the alignment. As shown in figure 1.22, for a detector with an effective detection surface and the normal of the surface with an angle of θ to the Lambertian surface, the irradiance becomes

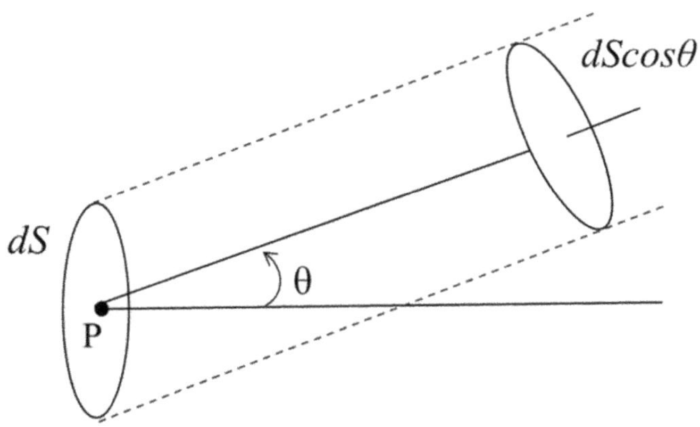

Figure 1.20. Measurement of intensity of an extended light source at a specific viewing angle.

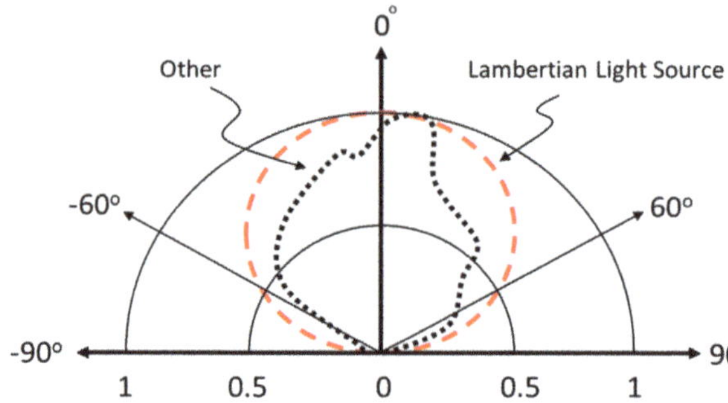

Figure 1.21. The intensity of a Lambertian light source is a circle in the radar figure.

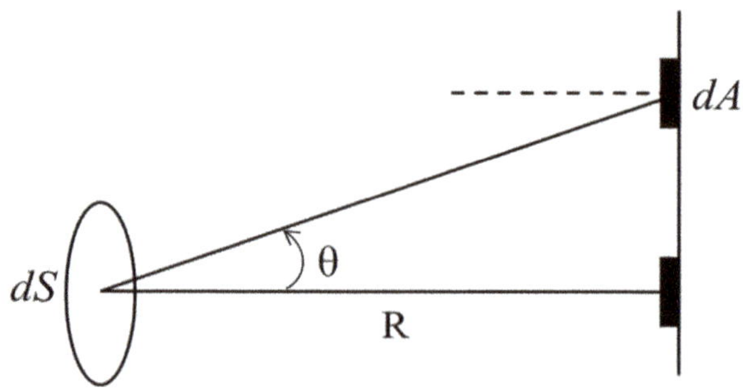

Figure 1.22. The normal direction of the irradiated surface does not point to the extended light source.

$$E_\theta = E_0 \cos^4 \theta. \tag{1.10}$$

This means that an extended light source (or surface) is more directional than a point source. To detect the flux emitted by a Lambertian light source is interesting and important. As shown in figure 1.23, with a light cone of θ_1, the total flux of a Lambertian light source can be written

$$F = \int dF = \pi B dS \sin^2 \theta_1. \tag{1.11}$$

For a whole field when θ_1 is extended to 90°, the total flux on a hemisphere can be written

$$F = \pi B dS \equiv M dS. \tag{1.12}$$

Hence,

$$M = \pi B, \tag{1.13}$$

where M is the exiting flux density of the light source and is called exitance. Equation (1.13) is especially useful to an engineer for estimating the total flux once the light source or surface performs Lambertian-like behavior. If the light source area is known and irradiance is measurable, one can calculate radiance and finally estimate the total flux of the light source using equation (1.13).

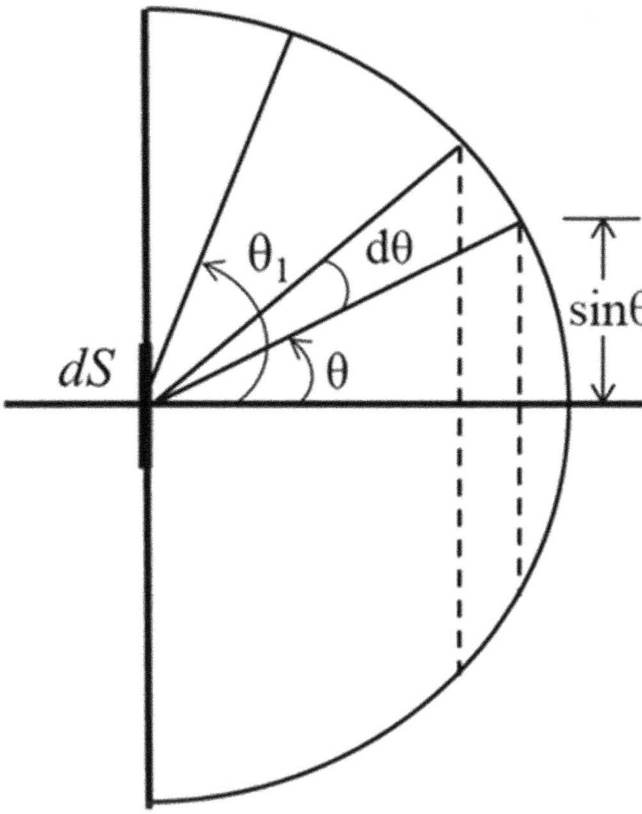

Figure 1.23. A Lambertian light source emits light to a hemisphere space.

To consider the flux transfer from one to another, the equation can be expressed

$$dF = B(x, y, L, M)dxdydLdM, \tag{1.14}$$

where x and y are the coordinates in the first plane, where L and M are the ray vector projections along x and y axis, respectively; x′ and y′ are the coordinates in the second plane, where L' and M' are the ray vector projections along x' and y' axis, respectively. The reverse transfer is written

$$dF' = B'(x', y', L', M')dx'dy'dL'dM'. \tag{1.15}$$

If the flux transfer is reversible and lossless, $dF = dF'$, and we can write

$$B(x, y, L, M)dxdydLdM = B'(x', y', L', M')dx'dy'dL'dM'. \tag{1.16}$$

Equation (1.16) can be derived to another form

$$\frac{B}{n^2} = \frac{B'}{n'^2}, \tag{1.17}$$

where n and n' are the refractive indices in the first and second spaces, respectively. B/n^2 is an invariant quantity from one space to another through the refraction on an interface, and it is called radiance theorem.

In lighting optics, most of the detectors are human eyes. The sensation of brightness is an important factor to judge if the illumination is appropriate. To simulate human-eye response, the responsibility of the detector is needed to fit the sensation of human eyes. There are two photoreceptor cells in the retina, including rod and cone cells, which are able to detect visible light and transform to brightness and color information through the brain process. The responsibility by the cone and rod cells is shown in figure 1.24, where the response by the cone cells is called photopic vision, and by the rod cells is called scotopic vision [38]. The rods are responsible for vision at low light levels, do not mediate color vision and have a low spatial acuity. In contrast, cones are responsible at higher light levels with more color vision and have high spatial acuity. The light level range when the photopic and scotopic visions are operational simultaneously is called mesopic. The brightness sensed by the human eyes is in the unit of lumen in photometry instead of watt in radiometry. Thus to all the quantities in photometry are added 'luminous' to distinguish them from those in radiometry. The peak values κ of photopic and scotopic are 683 lm W^{-1} at the wavelength 555 nm, and 1754 lm W^{-1} at 507 nm, respectively. The parameter $V(\lambda)$ in figure 1.24 represents the relative spectral response of the human eye, and is a key factor to calculate the brightness, luminous flux Φ_l by the human eyes

$$\Phi_l = \kappa \int F(\lambda)V(\lambda)d\lambda, \tag{1.18}$$

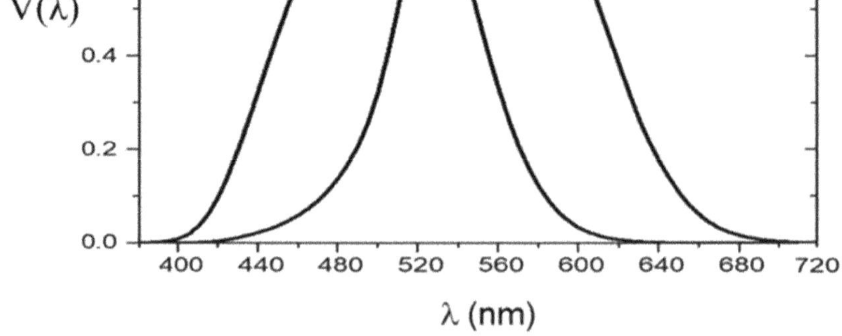

Figure 1.24. The spectral response of human eyes in photopic and scotopic visions.

Table 1.1. A list of various radiometric/photometric quantities.

Symbol	Radiometry	(Unit)	Photometry	(Unit)
F/Φ	Flux	Watt (W)	Luminous flux	Lumen (Lm)
I	Intensity	W sr^{-1}	Luminous intensity	Lm sr^{-1} (Candela, cd)
B	Radiance	W m^{-2}sr	Luminance	Lm m^{-2}sr (cd m^{-2} = nit)
E	Irradiance	W m^{-2}	Illuminance	Lm m^{-2} (lx or lux)
M	Radiant exitance	W m^{-2}	Luminous exitance	Lm m^{-2}

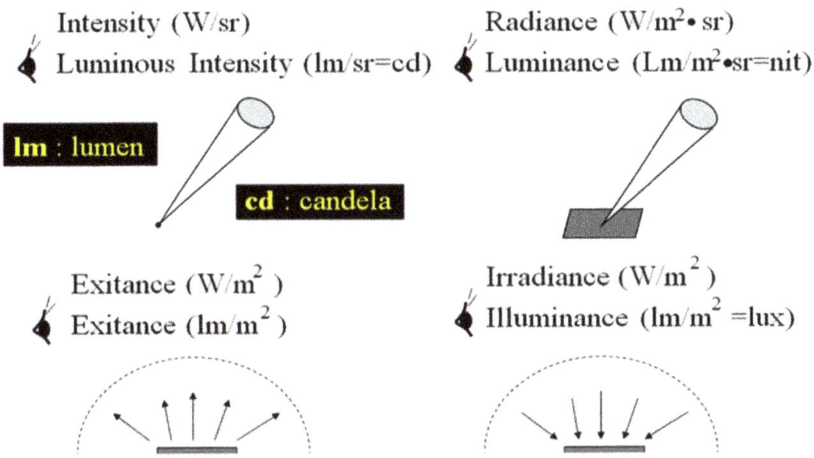

Figure 1.25. A summary of the characteristics of various radiometric/photometric quantities.

where $F(\lambda)$ is the radiometry flux at a specific wavelength. For example, a 1-watt laser light at 555 nm will perform 683 lm, but at 650 nm will perform 683 lm × 0.107 = 73.08 lm. A mixed light with 0.5 W at 555 nm and 0.5 W at 650 nm will perform 378.04 lm. With no doubt, the maximum brightness at 1W is 683 lm. When light is composed of various wavelengths, the 1 W brightness will be far below 683 lm. This is just the factor considering human eye response. In a solid-state light source, various factors including IQE, LEE and packaging efficiency could degrade the efficiency, so two thirds of the injection electric power could transfer to optical flux at most. Therefore, the total luminous efficacy could not exceed 300 lm W^{-1} [39, 40]. Practically, a white LED with luminous efficacy higher than 100 lm W^{-1} is regarded as high-level performance. Table 1.1 lists a comparison of the quantities and units between radiometry and photometry. Figure 1.25 clearly indicates the characteristics of various radiometric and photometric quantities.

1.6 Four-level optical designs and considerations

Solid-state lighting is a novel technology utilizing solid-state light sources for general and special lighting. The technology does not only concern the light source but also the illumination on the target. Thus there are several levels in handling the overall technology, from an LED die, packaging, a light source module to a lighting luminaire. From the viewpoint of element level to system level, optical approaches in solid-state lighting can be divided into four levels, which will be introduced in detail through this book.

The first level of optical design concerns the light property in LED dies. The problem does not touch the active quantum effect, but the passive LEE in the LED die. The injection electrons and holes recombine in the active layer, and light is emitted. If the emission of each location is tiny and the emission is isotropic, each point in the active layer can be regarded as a perfect point source. Since the active layer is very thin, the point sources are distributed across a thin layer, which can be regarded as a Lambertian light source on the double sides, i.e., face up and face down, as shown in figure 1.26. The problem arises because the light emission happens in a crystal with multiple layers, where the refractive indices are always larger than 1.7, which is the refractive index of the sapphire substrate. This will cause total internal refraction. It means that a ray along an angle larger than the critical angle will be reflected back to the LED die volume. Such an effect results in low LEE. The critical angle is around 24°, when the refractive index of the die volume is 2.5. If the LED die shape is a cuboid, only 4.3% of the emission light can be extracted from the LED die to the air. This problem is the zero-level of optics in solid-state lighting. To solve this problem, there are various approaches. One is to shape the LED die to avoid multiple total internal reflections in the die. Another is to encapsulate the LED die with a transparent medium, and the refractive index is between the LED die and the air. The other is to introduce microstructures on the top surface or the interface, to change the propagation direction when the light is reflected back to the LED die volume [29].

The active layer of an LED die is usually designed to emit light in a single color rather than white light. The most effective way to make up white light is to cover a phosphor layer/volume on a blue LED die. If the phosphor can emit yellow light, it is possible to form white light located at the black body radiation curve, through choosing an appropriate emission spectrum. The yellow-based phosphor, e.g., YAG phosphor, is the most convenient way to transfer blue light from a blue die to white light. The simplest way is to cover the blue die with the yellow phosphor, but the most difficult is to accurately control the CCT. Also, it is not easy to model this

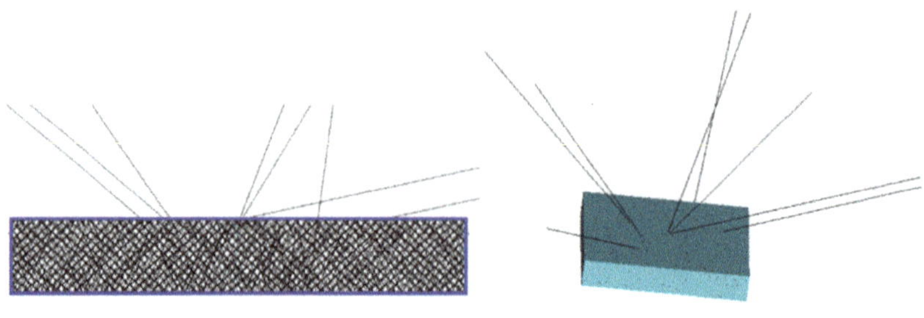

(a) (b)

Figure 1.26. Rays are trapped in the LED die volume. (a) Side view and (b) top view.

emission element. The phosphor acts in two roles in a white LED. One is to scatter the incoming blue light, and the second is to transfer the blue light to yellow light, as shown in figure 1.27. Mie scattering is the main mechanism of the scattering in phosphor. If the absorption coefficient and the quantum efficiency of the phosphor are known, the ratio between the yellow and the blue light can be calculated. Then the phosphor behavior could be successfully modelled. A white LED can be regarded as a light source with an appropriate packaging process. In the packaging structure, an LED die is bonded on a support metal plate with an electric circuit and pads, and covered with an encapsulant medium. Thus, calculation with a phosphor model incorporated with the packaging structure can figure out the light source behavior, and this is called light source model. A precise light source model is necessary when using a white LED for precise lighting design. To monitor the light source is the first level for solid-state lighting.

The second level of solid-state lighting is the optical design to perform desired illumination based on the precise light source model. In general, a white LED is a compact light source with light emission to a hemisphere space. This is different from the traditional light sources so that the optical design with an LED is in a different way. In the second-level design such as a spotlight, using a hybrid structure with a lens and reflector is an effective way to collect all the light emitted by the white LED. But this costs too much. Therefore, a new structure called TIR lens is proposed, where TIR means total internal reflection [41]. A TIR lens is a solid volume containing a small lens on the top face and a cone-shape reflection surface on the sidewall shown in figure 1.28. The lens and the reflector are bonded together to one element. In addition to a lens and

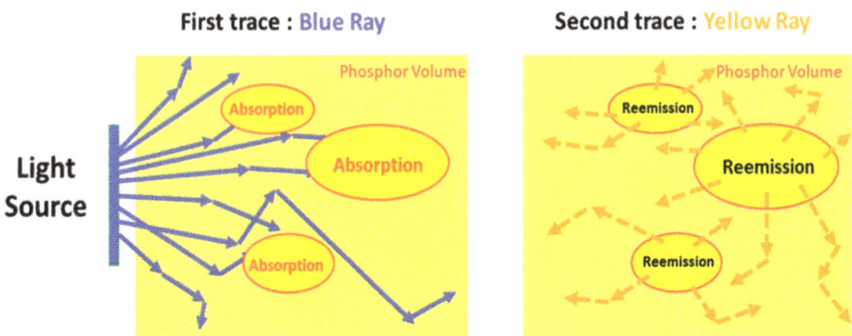

Figure 1.27. A schematic diagram of the mechanism of the yellow phosphor with a blue die.

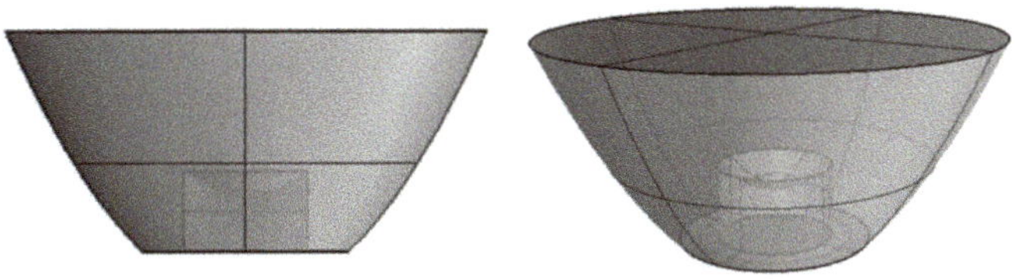

Figure 1.28. The structure of a TIR lens.

a reflector, there are various optical components which can be used in solid-state lighting with white LEDs. A diffuser can be used to enlarge the exit face area and reduce luminance. A novel diffuser with surface microstructure can be used to spread the light into a specific light pattern, as shown in figure 1.29. A multi-segmented reflector could be useful to spread the light from a white LED to make up a specific pattern, including a high-contrast cutoff line to meet the regulation of a headlamp [42]. As shown in figure 1.30, a freeform lens could be useful to redirect the light from a

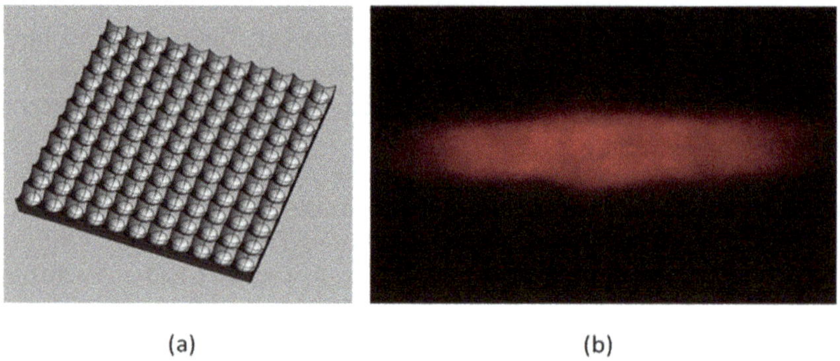

Figure 1.29. (a) Micro lens array serving a diffuser, and (b) the light pattern when illuminated with a red laser. Reprinted with permission from [43].

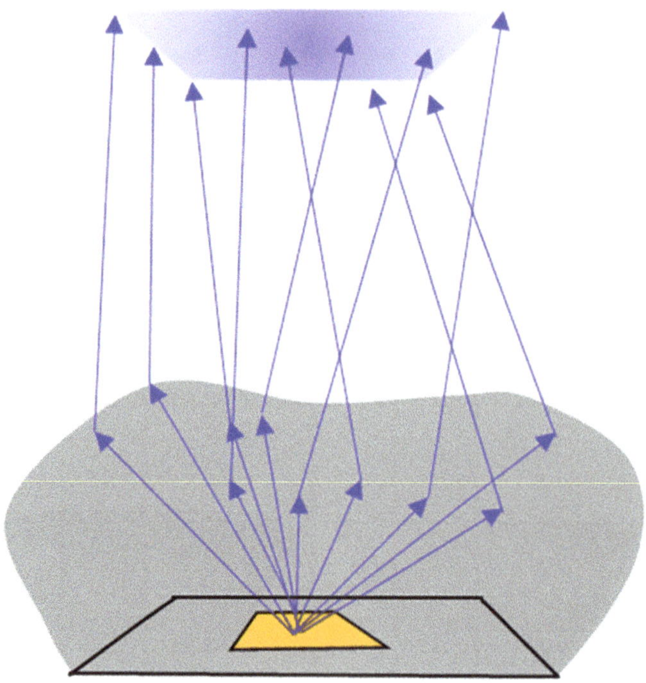

Figure 1.30. Schematic diagram of a freeform lens for beam shaping.

white LED to form an illumination light pattern such as a circle, a square or any desired shape. However, a designer must understand that a white LED is not a point source, especially when the optical component is not located at the far field from the light source. Thus, a precise light source model is necessary. In addition, a simulation tool with Monte Carlo ray tracing is very helpful in simulation with sufficient ray number to obtain an accurate result.

The third level of optical design is mainly aimed to prevent glare. Glare is a negative effect when the light affects a user's vision. A common condition is to project light into the user's eyes or the light source is too bright to disturb a user's vision. To avoid that, there are several approaches. The first is to reduce the luminance of the light source or the exit face. Here a diffuser is a good choice. A diffuser can redirect one propagation direction to a different one, which is randomly distributed. Through an appropriate design, the luminance can be effectively reduced with use of a diffuser or multiple diffusers. Another approach is especially useful in designing a headlamp. A headlamp projects a light pattern with a clear cutoff line, as shown in figure 1.31. Below the cutoff line is the bright zone for target illumination. Above the cutoff line is the dark zone for anti-glare. To make sure of the darkness in the dark zone, a multi-segment reflector is applicable. Each segment has a specific tilt angle to project light to the illumination area, and avoid the dark area, as shown in figure 1.32. A different approach is to use an imaging lens to project the light pattern, but a baffle is needed to block the light into the dark zone.

Although the four levels of optical design are all needed in a high-quality lighting design with a solid-state light source, a designer is not requested to be familiar with all the four-level techniques. The zero level is for an engineer in chip processing, the first level is for a packaging engineer; the first, second and third levels are for an optical designer, who can not only utilize all the useful optical components to project light into the illumination area, but also can avoid projecting light to the dark area or reduce the luminance of all the light sources or possible reflection surfaces to

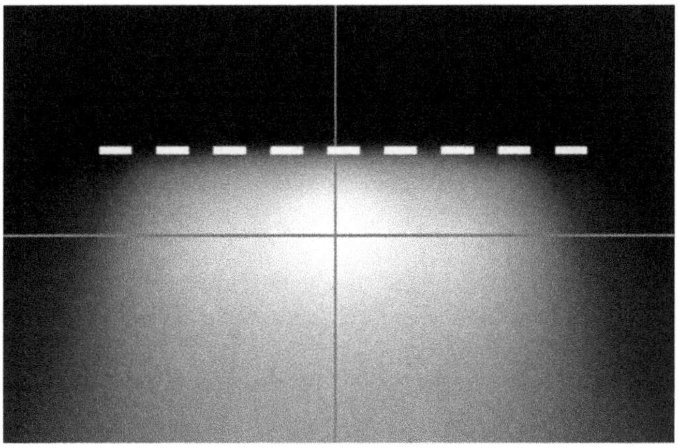

Figure 1.31. A light pattern with a high-contrast cutoff line (dashed yellow line).

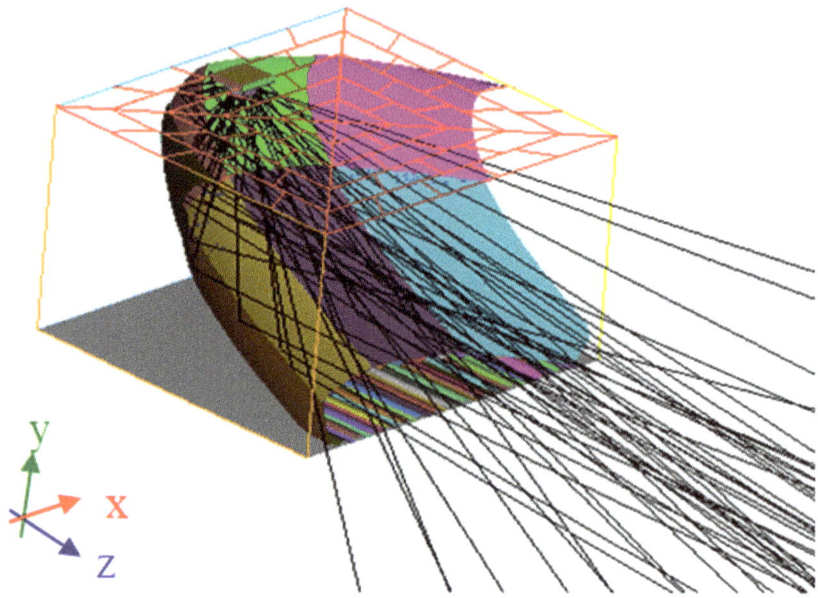

Figure 1.32. A multi-segment reflector can be designed to form a cutoff line as shown in figure 1.31. Reprinted with permission from [44].

induce glare. Using the all-level techniques is useful to design an effective lighting luminaire with high energy efficiency and to build up a comfortable lighting environment.

References

[1] Biard J R and Pittman G E 1966 Semiconductor radiant diode *US Patent* 3,293,513
[2] Nick H J and Bevacqua S F 1962 Coherent (visible) light emission from Ga($As_{1-x}P_x$) junctions *Appl. Phys. Lett.* **1** 82–3
[3] Sugawara H, Ishikawa M and Hatakoshi G 1991 High-efficiency InGaAlP/GaAs visible light-emitting diodes *Appl. Phys. Lett.* **58** 1010–2
[4] Nakamura S, Senoh M and Mukai T 1993 P-GaN/N-InGaN/N-GaN double-heterostructure blue-light-emitting diodes *Jpn. J. Appl. Phys.* **32** L8–11
[5] Shimizu Y, Sakano K, Noguchi Y and Moriguchi T 1999 Light emitting device having a nitride compound semiconductor and a phosphor containing a garnet fluorescent material *US Patent* 5,998,925
[6] Schubert E F and Kim J K 2005 Solid-state light sources getting smart *Science* **308** 1274–8
[7] Jones E D 2001 *Light Emitting Diodes (LEDs) for General Illumination* (Washington: OIDA Publishing)
[8] Žukauskas A, Shur M S and Gaska R 2002 *Introduction to Solid-state Lighting* (New York: Wiley)
[9] Karlicek R, Sun C C, Zissis G and Ma R 2016 *Handbook of Advanced Lighting Technology* (Berlin: Springer Int. Publishing)
[10] Narukawa Y 2004 White-light LEDs *Opt. Photonics News* **15** 24–9

[11] Cho J, Park J H, Kim J K and Schubert E F 2017 White light-emitting diodes: history, progress, and future *Laser Photonics Rev.* **11** 1600147
[12] Liu Z *et al* 2020 Micro-light-emitting diodes with quantum dots in display technology *Light: Sci. Appl.* **9** 83
[13] Mertens R 2020 *The Micro LED Handbook* (Morrisville: Lulu Press, Inc. Publishing)
[14] Jiang H X and Lin J Y 2013 Nitride micro-LEDs and beyond—a decade progress review *Opt. Express* **21** A475–84
[15] Wu T, Sher C W, Lin Y, Lee C F, Liang S, Lu Y, Chen S W, Guo W, Kuo H C and Chen Z 2018 Mini-LED and micro-LED: promising candidates for the next generation display technology *Appl. Sci.* **8** 1557
[16] Wu Y, Ma J, Su P, Zhang L and Xia B 2020 Full-color realization of micro-LED displays *Nanomaterials* **10** 2482
[17] Huang Y, Tan G, Gou F, Li M C, Lee S L and Wu S T 2019 Prospects and challenges of mini-LED and micro-LED displays *J. Soc. Inf. Disp.* **27** 387–401
[18] Sun C C, Chen C Y, Chen C C, Chiu C Y, Peng Y N, Wang Y H, Yang T H, Chung T Y and Chung C Y 2012 High uniformity in angular correlated-color-temperature distribution of white LEDs from 2800 K to 6500 K *Opt. Express* **20** 6622–30
[19] Speier I and Salsbury M 2006 Color temperature tunable white light LED system *Proc. SPIE* **6337** 63371F
[20] He G, Zheng L and Yan H 2010 LED white lights with high CRI and high luminous efficacy *Proc. SPIE* **7852** 78520A
[21] Yadav P J, Joshi C P and Moharil S V 2013 Two phosphor converted white LED with improved CRI *J. Lumin.* **136** 1–4
[22] Kuwano Y, Kaga M, Morita T, Yamashita K, Yagi K, Iwaya M, Takeuchi T, Kamiyama S and Akasaki I 2013 Lateral hydrogen diffusion at p-GaN layers in nitride-based light emitting diodes with tunnel junctions *Jpn. J. Appl. Phys.* **52** 08JK12
[23] Jeon S R, Song Y H, Jang H J, Yang G M, Hwang S W and Son S J 2001 Lateral current spreading in GaN-based light-emitting diodes utilizing tunnel contact junctions *Appl. Phys. Lett.* **78** 3265–7
[24] Schubert E F 2018 *Light-emitting Diodes* (Cambridge: Cambridge University Press Publishing)
[25] Svensson C T, Mårtensson T, Trägårdh J, Larsson C, Rask M, Hessman D, Samuelson L and Ohlsson J 2008 Monolithic GaAs/InGaP nanowire light emitting diodes on silicon *Nanotechnology* **19** 305201
[26] Chen C H, Hargis M, Woodall J M, Melloch M R, Reynolds J S, Yablonovitch E and Wang W 1999 GHz bandwidth GaAs light-emitting diodes *Appl. Phys. Lett.* **74** 3140–2
[27] Liu M, Rong B and Salemink H W 2007 Evaluation of LED application in general lighting *Opt. Eng.* **46** 074002
[28] Jeong S M, Kissinger S, Kim D W, Lee S J, Kim J S, Ahn H K and Lee C R 2010 Characteristic enhancement of the blue LED chip by the growth and fabrication on patterned sapphire (0 0 0 1) substrate *J. Cryst. Growth* **312** 258–62
[29] Lee T X, Gao K F, Chien W T and Sun C C 2007 Light extraction analysis of GaN-based light-emitting diodes with surface texture and/or patterned substrate *Opt. Express* **15** 6670–6
[30] Hsu S C and Liu C Y 2006 Fabrication of thin-GaN LED structures by Au–Si wafer bonding *Electrochem. Solid-State Lett.* **9** G171
[31] Ha M and Graham S 2012 Development of a thermal resistance model for chip-on-board packaging of high power LED arrays *Microelectron. Reliab.* **52** 836–44

[32] Kim L and Shin M W 2007 Thermal resistance measurement of LED package with multichips *IEEE Trans. Compon. Packag.* **30** 632–6

[33] Mertol A 2000 Application of the Taguchi method to chip scale package (CSP) design *IEEE Trans. Adv. Packag.* **23** 266–76

[34] Zhang T, Tang H, Li S, Wen Z, Xiao X, Zhang Y, Wang F, Wang K and Wu D 2017 Highly efficient chip-scale package LED based on surface patterning *IEEE Photon. Technol. Lett.* **29** 1703–6

[35] Huang C H, Chen K J, Tsai M T, Shih M H, Sun C W, Lee W I, Lin C C and Kuo H C 2015 High-efficiency and low assembly-dependent chip-scale package for white light-emitting diodes *J. Photon. Energy* **5** 057606

[36] McCluney W R 2014 *Introduction to Radiometry and Photometry* (Boston, MA: Artech House)

[37] Palmer J M and Grant B G 2010 *The Art of Radiometry* (Bellingham, WA: SPIE Press)

[38] Mahajan V N 1998 *Optical Imaging and Aberrations: Part I. Ray Geometrical Optics* (Bellingham, WA: SPIE Press)

[39] Narukawa Y, Ichikawa M, Sanga D, Sano M and Mukai T 2010 White light emitting diodes with super-high luminous efficacy *Appl. Phys.* **43** 354002

[40] Murphy T W Jr 2012 Maximum spectral luminous efficacy of white light *J. Appl. Phys.* **111** 104909

[41] Hecht E 2002 *Optics* (Reading, MA: Wesley)

[42] Sun C C, Wu C S, Hsieh C Y, Lee Y H, Lin S K, Lee T X, Yang T H and Yu Y W 2021 Single reflector design for integrated low/high beam meeting multiple regulations with light field management *Opt. Express* **29** 18865–75

[43] Lee X H 2012 Design and manufacture of surface-structured diffuser and the applications *PhD Thesis* National Central University

[44] Cai J Y 2015 A study on illumination light pattern of high performance and multi-functional LED projection lamps *PhD Thesis* National Central University

IOP Publishing

Optical Design for LED Solid-State Lighting

A guide

Ching-Cherng Sun and Tsung-Xian Lee

Chapter 2

Basic optics

This chapter briefly introduces the fundamental principles of optics. Its primary objective is to provide readers with the optical terminology and background knowledge needed for this book. Readers who want to learn more about optics can further read the references of this chapter.

2.1 Propagation of light

Any kind of traveling wave at a velocity v satisfies the wave equation

$$\frac{\partial^2 W(r, t)}{\partial r^2} = \frac{1}{v^2}\frac{\partial^2 W(r, t)}{\partial t^2}, \tag{2.1}$$

where r is the spatial dependence and t is the temporal dependence of the wave function. One of the solutions of equation (2.1) is a sinusoidal function. With regard to a light wave, it travels from one point to another according to *Fermat's principle*, where a light wave travels in a path between two points at the shortest time [1]. In general, as shown in figure 2.1, we can easily express the light wave in a sinusoidal form

$$W(r, t) = A\cos\left\{2\pi\left(ft - \frac{r}{\lambda}\right) + \varphi_i\right\} = A\cos\{\varpi t - \bar{k}\cdot\bar{r} + \varphi_i\}, \tag{2.2}$$

where A is the amplitude, f is the frequency, ϖ is the angular frequency, λ is the wavelength, $\bar{r}$ is the position vector, $\bar{k}$ is the wave vector, which is along the wavefront normal, and φ_i is the initial phase of the light wave, which is related to temporal coherence, and will not be shown in the following discussions for simplicity. In an isotropic medium, the wave vector coincides with the propagation direction of light, which represents the ray direction and can be expressed by the Poynting vector. This condition does not hold as in an anisotropic medium. Fortunately, most mediums are isotropic. Thus we can generally regard the wave-front normal, i.e., the direction of the wave vector, as the propagation direction of

doi:10.1088/978-0-7503-2368-0ch2

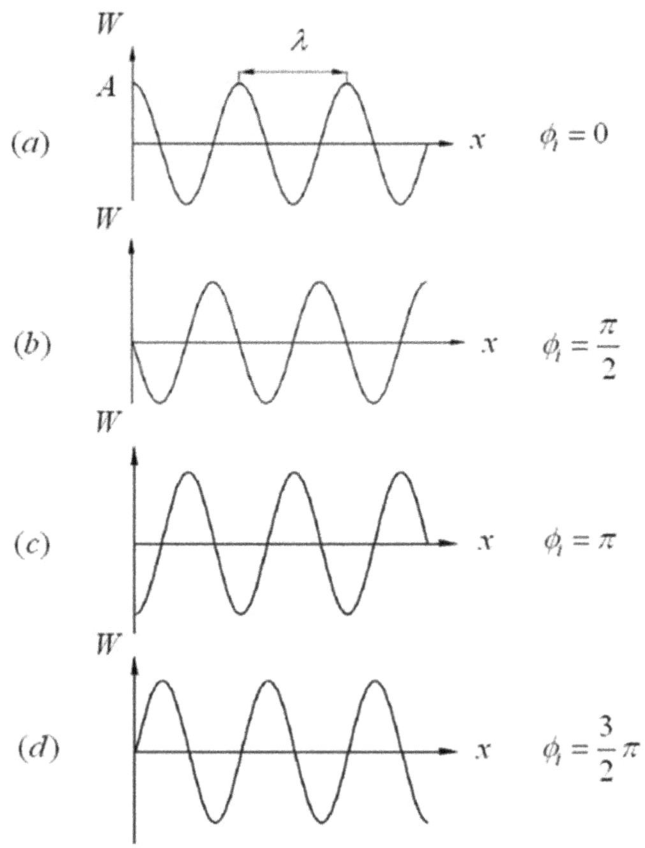

Figure 2.1. Sinusoidal waves with different initial phases.

the light wave. Therefore, it is so simple to draw an arrow to express the direction of propagation. When the wavefront normal is not unique, there are various wave vectors, as shown in figure 2.2. Using such a simple technique, we can easily figure out image formation and locate the light propagation direction through different isotropic media. This is called ray optics.

2.2 Complex representation of light waves

For convenience in calculation, it is general to use a complex expression to replace the sinusoidal wave of light. Thus, equation (2.2) can be rewritten as

$$W(r, t) = \mathrm{Re}\{Ae^{i(\varpi t - \bar{k}\cdot\bar{r})}\} = \mathrm{Re}\{E(r, t)\}. \tag{2.3}$$

If there are two light waves $P(r,t)$ and $Q(r,t)$, which are expressed as

$$W_1(r, t) = A_1 \cos(w_1 t - k_1 \cdot r) = \mathrm{Re}\{E_1(r, t)\}, \tag{2.4}$$

$$W_2(r, t) = A_2 \cos(w_2 t - k_2 \cdot r) = \mathrm{Re}\{E_2(r, t)\}, \tag{2.5}$$

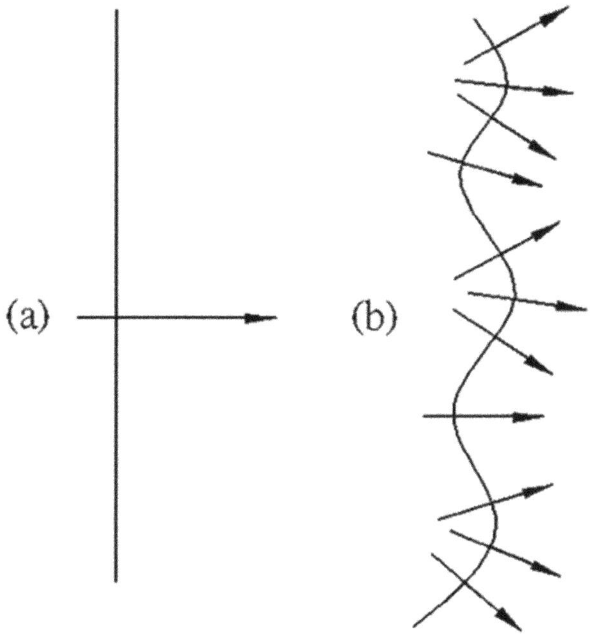

Figure 2.2. Examples of wavefront normal.

where

$$E_1(r,\ t) = A_1 e^{i(w_1 t - k_1 \cdot r)}, \tag{2.6}$$

$$E_2(r,\ t) = A_2 e^{i(w_2 t - k_2 \cdot r)}, \tag{2.7}$$

we can easily find that

$$W_1(r,\ t)W_2(r,\ t) \neq \mathrm{Re}\{E_1(r,\ t)E_2(r,\ t)\}. \tag{2.8}$$

It is generally true that the product of two real parts of the complex function does not equal the real part of the product of the two complex functions.

In a light wave, the electric field oscillates at a frequency as high as 600 THz when the wavelength is 0.5 μm. Therefore, in regard to the energy intensity of a light wave, time-averaged values rather than instantaneous values are often considered. In general, we have to find the time-averaged value of the product of two sinusoidal functions, which can be written

$$< W_1(r,\ t)W_2(r,\ t)> = \frac{1}{T}\int_0^T W_1(r,\ t)W_2(r,\ t)dt, \tag{2.9}$$

where T is a period much longer than the oscillation of the light wave. When the frequency of both light waves is equal to each other, we obtain

$$< W_1(r,\ t)W_2(r,\ t)> = \frac{1}{2}A_1A_2 \cos\{(k_1 - k_2)\cdot r\}. \tag{2.10}$$

By use of the complex expression shown in equations (2.4)–(2.7), we find that equation (2.10) can be rewritten

$$<W_1(r,\, t)W_2(r,\, t)> = \frac{1}{2}\,\mathrm{Re}\{E_1(r,\, t)E_2^*(r,\, t)\}, \tag{2.11}$$

where the asterisk indicates the complex conjugate. Based on electromagnetic theory, we obtain the time-average power density of a light wave

$$\langle I(r,\, t)\rangle = v\varepsilon_m\langle E^2(r,\, t)\rangle = \frac{v\varepsilon_m}{2}\,|E|^2 = \frac{v\varepsilon_m}{2}A^2, \tag{2.12}$$

where ε_m is known as the electric permittivity of the medium. In this book, we drop the time average symbolism with it being understood that the power density is always the time average. For further simplicity, the light intensity is expressed

$$I = |E|^2. \tag{2.13}$$

2.3 Huygens' principle

Huygens' principle is one of the most significant principles in wave optics, especially in diffraction [2]. In principle, by using graphs, it is possible to obtain the shape of a wavefront at any instant when given a wavefront of an earlier instant. The principle essentially states that every point of the wavefront can be regarded as the new point source emitting spherical wavelets forward. Thus a new wavefront can be constructed graphically by drawing a surface tangent to all the secondary spherical wavelets. If the propagation velocity is not constant at all parts of the wavefront, each spherical wavelet should be drawn with an appropriate wave velocity.

Figure 2.3 gives an illustrated example for the use of Huygens' principle. The known wavefront is shown as the surface Σ, called the primary wavefront, and the directions of wave propagation are indicated with small arrows. To determine the wavefront at an interval of time Δt with a wave velocity v, we construct a series of spheres (i.e., the secondary spherical wavelets) of radius $r = v\Delta t$ from each point of surface wavefront Σ. These spheres represent the secondary wavelets. Now if we draw a surface tangent to all the surfaces of the spheres, we get the shape of a new wavefront, called the secondary wavefront. We note that the wave velocities at every

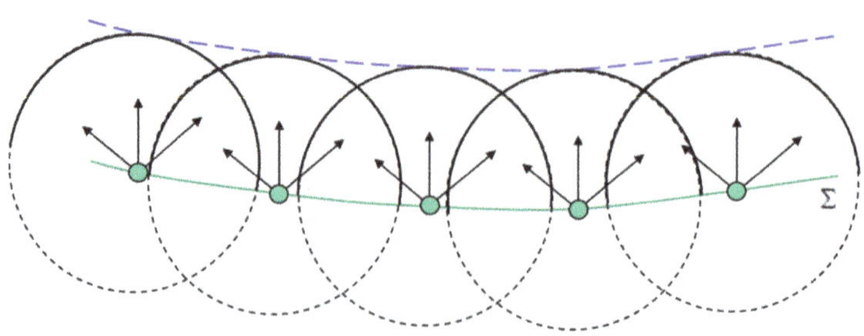

Figure 2.3. A secondary wavefront is formed as described in Huygens' principle.

point on the surface are assumed to be equal. This is a typical example of light wave propagation in a homogeneous isotropic medium.

Furthermore, according to Huygens' principle, a backward wavefront should form the secondary spherical wavelets. This phenomenon has never been observed, however. The discrepancy can be explained by the application of Huygens' principle in predicting a new wavefront. Owing to Huygens' principle, the wave nature of light propagation becomes more fully understood.

2.4 The speed of light

The optical wave is an electromagnetic field, which can be described by Maxwell's equations:

$$\nabla \times E + \frac{\partial B}{\partial t} = 0, \tag{2.14}$$

$$\nabla \times H - \frac{\partial D}{\partial t} = J, \tag{2.15}$$

$$\nabla \cdot D = \rho, \tag{2.16}$$

$$\nabla \cdot B = 0, \tag{2.17}$$

where E and H are called the electric field and the magnetic field, respectively; B and D are called the electric displacement and the magnetic induction; J is the electric current density and ρ is the electric charge density. The current density and the charge density may be regarded as the source of the electromagnetic radiation. Both of them are zero when a light wave is far from the sources. To consider the matter response to the light field, the electric displacement and the magnetic induction can be expressed as

$$D = \varepsilon E = \varepsilon_0 E + P, \tag{2.18}$$

$$B = \mu H = \mu_0 H + M, \tag{2.19}$$

where ε and μ are the dielectric tensor and the permeability tensor, respectively; ε_0 and μ_0 are the permittivity and permeability of vacuum, respectively; and P and M are the electric and magnetic polarizations, respectively.

When a light wave is propagated in a free space and is far from the source, the electric displacement and the magnetic induction are proportional to the electric field and the magnetic field of the light wave, respectively. Then we can easily obtain a wave equation from equations (2.14)–(2.19),

$$\frac{\partial^2 E(r, t)}{\partial r^2} = \varepsilon_0 \mu_0 \frac{\partial^2 E(r, t)}{\partial t^2}. \tag{2.20}$$

In comparison with the general expression of the wave equation in equation (2.1), we obtain the speed of light in vacuum

$$v_0 = \frac{1}{\sqrt{\varepsilon_0 \mu_0}} \approx 3 \times 10^8 \text{ m s}^{-1}. \tag{2.21}$$

Similarly, the light speed in an optical medium is

$$v = \frac{1}{\sqrt{\varepsilon \mu}}. \tag{2.22}$$

The ratio of the light speed in vacuum to that in a medium is defined as the refractive index of the medium,

$$n = \frac{v_0}{v} = \sqrt{\frac{\varepsilon \mu}{\varepsilon_0 \mu_0}}. \tag{2.23}$$

The refractive index of a medium is larger than 1 because the light speed in vacuum is higher than that in an optical medium. When a light wave is traveling in an optical medium, the frequency keeps unchanged, but the wavelength becomes n^{-1} of that in vacuum. As a result, the accumulating phase of the light wave should be written as

$$\varphi = \frac{2\pi n d}{\lambda_0} = k_0 n d = k_0 (OPL), \tag{2.24}$$

where d is the geometrical path length as the light wave travels in the medium, λ_0 is the wavelength in vacuum, k_0 is the wave number in vacuum and OPL is defined as optical path length (so-called OPL), which is the product of the refractive index of the optical medium and the geometrical length that the light wave has passed through. In regards to optical problems, OPL rather than geometrical length is always an important parameter because it is proportional to the phase of a light wave.

2.5 Wavefront

A surface consisting of the points with equal OPL from the light source is defined as the wavefront of a light wave. Owing to the characteristic of the light source, the geometry of the wavefront may be different from one light wave to another, and different from one propagation distance to another. The wavefront structure will be changed once the light wave passes through a medium with a spatial-dependence refractive index. There are two essential wavefronts introduced below, i.e., a plane wave and a spherical wave [1, 3].

Plane waves should be the simplest wavefront structure, where the surface of constant OPL forms a plane, as shown in figure 2.4. The planar wavefront has a unique wave vector, and the surface of the constant OPL can be expressed

$$\bar{k} \cdot (\bar{r} - \bar{r}_0) = 0, \tag{2.25}$$

where $\bar{r}$ is the position vector and $\bar{r}_0$ is the given position vector points to a certain point located at the planar surface. Because the wave vector and the given position vector are well defined, equation (2.25) can be rewritten as

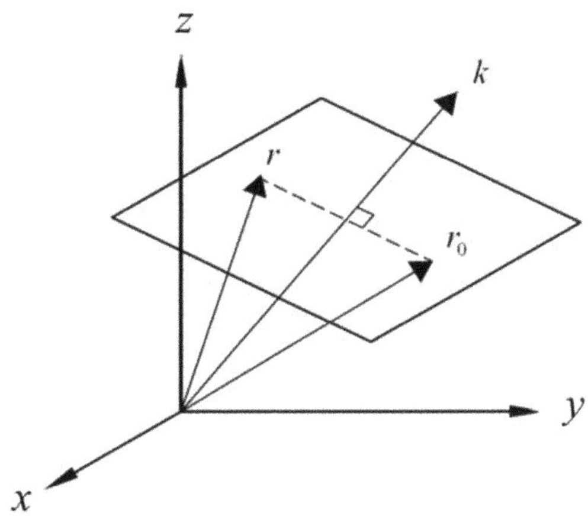

Figure 2.4. A plane wave.

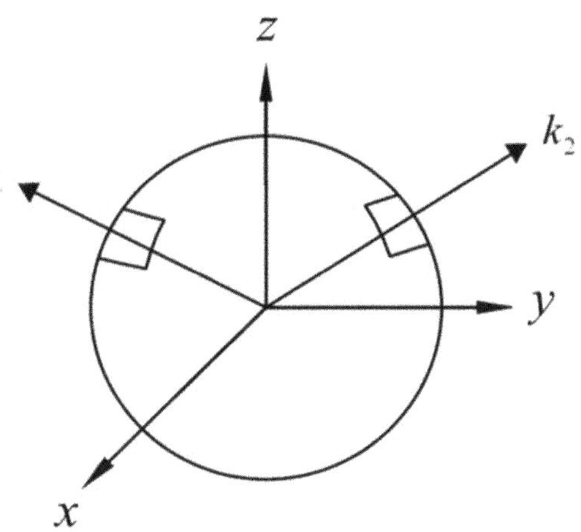

Figure 2.5. A spherical wave.

$$\bar{k} \cdot \bar{r} = \text{constant}. \tag{2.26}$$

Therefore, we can express the wave function of a plane wave

$$W(r, t) = A\cos\{\varpi t - \bar{k} \cdot \bar{r}\} \tag{2.27}$$

or

$$W(r, t) = Ae^{i(\varpi t - \bar{k} \cdot \bar{r})}. \tag{2.28}$$

Equations (2.27) and (2.28) are the simplest and basic expression of a light wave, for the frequency and wave propagation are both unique. Such a light wave is called a *monochromatic plane wave.*

A perfect spherical wave emerges from a point source. The wavefront for a constant OPL is a sphere centered by the point source, as shown in figure 2.5.

The wave vector is always parallel to the position vector, so the surface of constant OPL can be expressed as

$$kr = \text{constant}, \tag{2.29}$$

where $r = \sqrt{x^2 + y^2 + z^2}$. Therefore, we can express the spherical wave

$$W(r, t) = \frac{A}{r}\cos(\varpi t - kr) \tag{2.30}$$

or

$$W(r, t) = \frac{A}{r}e^{i(\varpi t - kr)}. \tag{2.31}$$

As shown in figure 2.6, when the observation plane is around the z-axis, we may express

$$r = z\sqrt{1 + \frac{x^2 + y^2}{z^2}} = z\left(1 + \frac{\theta^2}{2} - \frac{\theta^4}{8} + \ldots\right), \tag{2.32}$$

where $\theta = \left(\frac{x^2 + y^2}{z^2}\right)^{1/2}$. Once $\theta << 1$, i.e., in the paraxial condition, equation (2.32) can be written

$$r = z\left(1 + \frac{\theta^2}{2}\right). \tag{2.33}$$

Then the spherical wave is written as

$$W(r, t) = \frac{A}{z}e^{i(\varpi t - kz - k\frac{x^2+y^2}{z})}. \tag{2.34}$$

At an observation plane centered at $z = z_0$, the paraxial spherical wave is rewritten as

$$W(r, t) = A'e^{i(\varpi t - k\frac{x^2+y^2}{z_0})}, \tag{2.35}$$

where A' is a constant amplitude across the plane. The spatial-dependent phase term shown in equations (2.34) and (2.35) illustrates that the paraxial spherical wavefront is an elliptical wavefront.

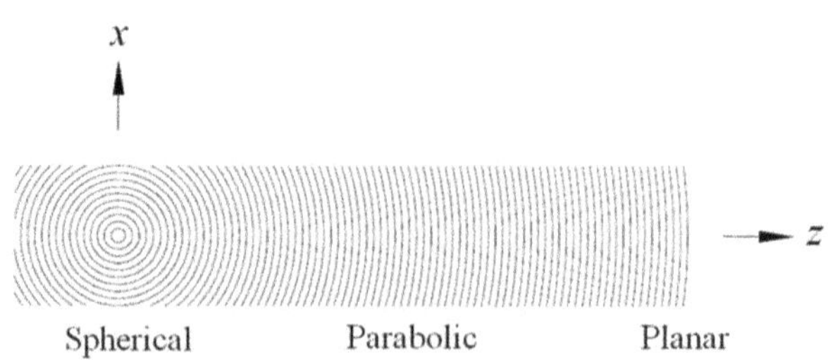

Figure 2.6. A spherical wave is propagated along the z-axis. The further the propagation distance is, the flatter the wavefront.

2.6 Polarization

Regarding the nature of electric and magnetic field oscillation, the oscillation status of both fields, i.e., the polarization state, should be indicated in a complete light wave expression [4, 5]. Let us assume that a monochromatic plane wave is propagating along the z-axis, as illustrated in figure 2.7, in which the electric and magnetic fields are oscillating along straight lines at right angles to one another, perpendicular to the direction of propagation, then the wave is said to be *linearly polarized* (in the x direction), which means that at any fixed point of the electric (or magnetic) field vector oscillates along a line perpendicular to the wave propagation direction. If the electric field does not oscillate along a straight line and changes the direction, such as a (clockwise) circle as with the wave propagation shown in figure 2.8(a), then the light wave is called clockwise circularly polarized light. On the other hand, if the electric vector describes a counterclockwise circle, as shown in figure 2.8(b), it is termed *counterclockwise circularly polarized light.* Furthermore, any electric field vector E can be decomposed into components E_x and E_y. If $E_x \neq E_y$, and assuming a constant phase shift between them, the resultant electric field vector E will follow an ellipse form, as shown in figure 2.9, where the light wave

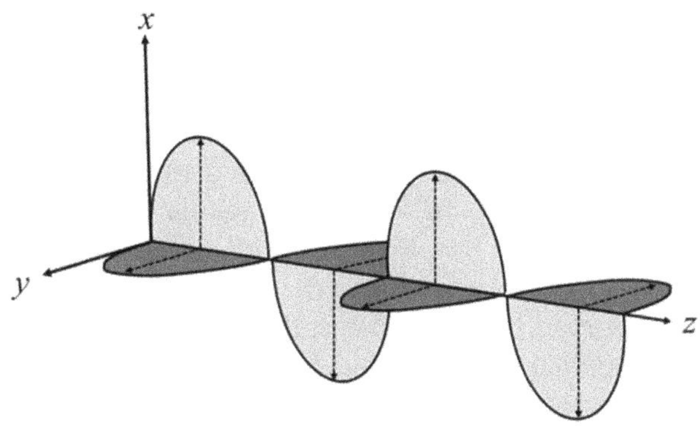

Figure 2.7. Monochromatic plane wave propagating along the z-axis.

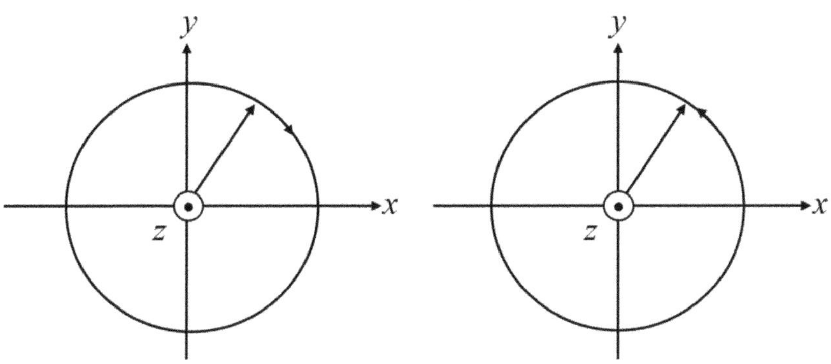

Figure 2.8. Circularly polarized light: (a) clockwise, and (b) counterclockwise.

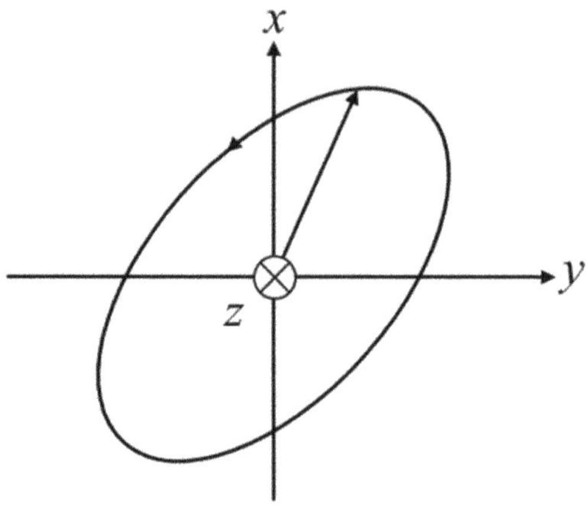

Figure 2.9. Elliptically polarized light.

is *elliptically polarized* light. Nevertheless, if the electric field vectors are aligned in a manner symmetrically distributed in all directions perpendicular to the propagation, the light field is known as *unpolarized* or *natural* light.

2.7 Fresnel equations of reflection

Maxwell equations are useful in determining the energy distribution of the light wave, especially through optical interfaces. When a monochromatic plane wave is incident on a planar boundary between two media which are assumed to be homogeneous, linear, isotropic, nondispersive and nonmagnetic, we can divide the incident light into two different types due to the boundary conditions associated with the electromagnetic theory. The first one is called TE case, where the electric field of the light wave is perpendicular to the incident plane, and the polarization of the incident light is called *s* polarization. The other one is TM case, where the electric field of the light wave is parallel to the plane of incidence, and the polarization of the incident light is called *p* polarization [2, 3]. The polarization of the *p* and *s* components are illustrated as bars and dots, respectively, in figure 2.10.

It has been observed in the past that the ratio of the *reflectance* to the *transmittance* depends not only on the incident angle and refractive indices of the media, but also on the polarization state of the incident beam into two orthogonal polarization components. Assuming that the refractive indices of the two media are n_1 and n_2, respectively, the reflection and refraction coefficients in the TE mode are given by

$$r_s = \frac{E_{2s}}{E_{0s}} = \frac{n_1 \cos\theta_0 - n_2 \cos\theta_1}{n_1 \cos\theta_0 + n_2 \cos\theta_1} = \frac{\sin(\theta_1 - \theta_0)}{\sin(\theta_1 + \theta_0)}, \tag{2.36}$$

$$t_s = \frac{E_{1s}}{E_{0s}} = 1 + r_s = \frac{2\sin\theta_1 \cos\theta_0}{\sin(\theta_1 + \theta_0)}, \tag{2.37}$$

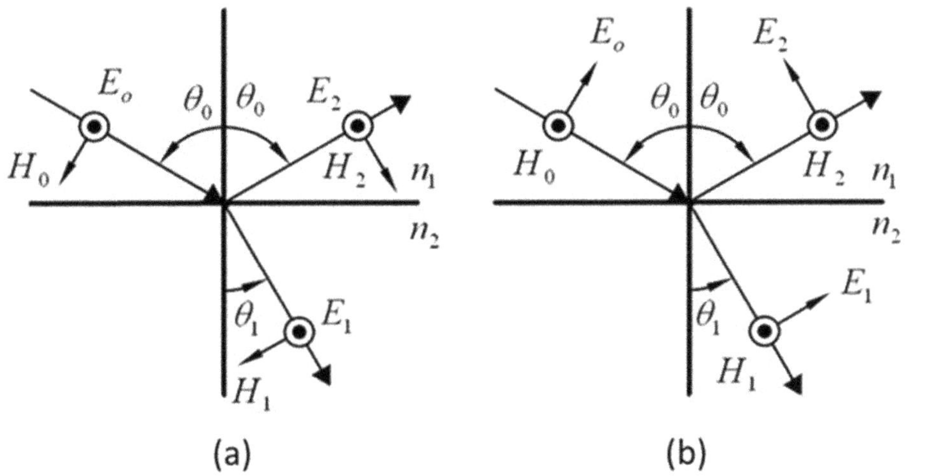

Figure 2.10. The incident, reflection and refraction rays on the incident plane in (a) the TE or p polarization, and (b) TM or s polarization.

and in the TM mode

$$r_p = \frac{E_{2p}}{E_{0p}} = \frac{n_2 \cos \theta_0 - n_1 \cos \theta_1}{n_2 \cos \theta_0 + n_1 \cos \theta_1} = \frac{\tan(\theta_1 - \theta_0)}{\tan(\theta_1 + \theta_0)}, \tag{2.38}$$

$$t_p = \frac{E_{1p}}{E_{0p}} = \frac{n_1}{n_2}(1 + r_p) = \frac{2 \sin \theta_1 \cos \theta_0}{\sin(\theta_1 + \theta_0)\cos(\theta_0 - \theta_1)}. \tag{2.39}$$

These equations are known as the *Fresnel equations* and the coefficients are called the Fresnel coefficients. The symbols E_{0S}, E_{1S}, E_{2S}, E_{0P}, E_{1P} and E_{2P} represent the incident, refracted and reflected electric vectors. Since the energy is proportional to the absolute square of the field amplitude, the *reflectance* of the boundary can be expressed as

$$R_s = |r_s|^2,$$

$$R_p = |r_p|^2. \tag{2.40}$$

For normal incidence, i.e., $\theta_0 = 0$, we find that R_s and R_p reduce the same value,

$$R_s = R_p = \left[\frac{n_1 - n_2}{n_1 + n_2}\right]^2. \tag{2.41}$$

Thus, the reflectance at normal incidence is 4% for $n_1 = 1.5$ (glass) and $n_2 = 1$, and is about 32% for $n_1 = 3.6$ (GaAs) and $n_2 = 1$. The magnitudes $|r_s|$ and $|r_p|$ are plotted as functions of the angle of incidence θ_0 in figure 2.11 for internal reflection ($n_1 > n_2$) and figure 2.12 for external reflection ($n_1 < n_2$), whereas we find two special angles for total reflection and total transmission, respectively.

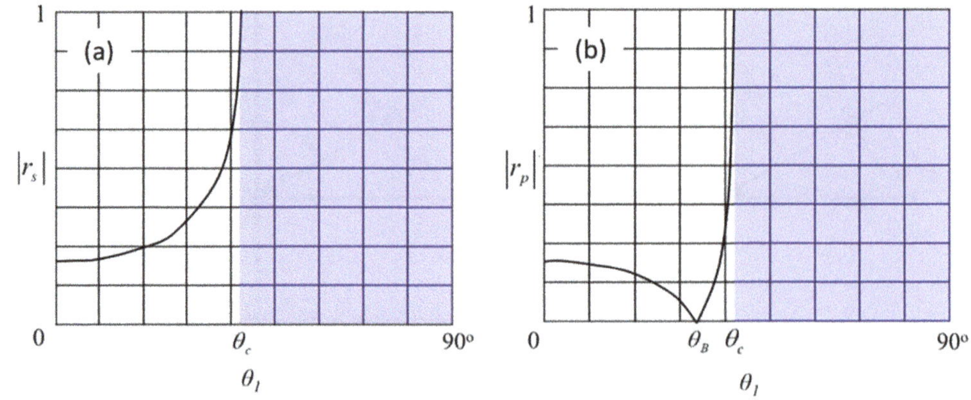

Figure 2.11. Reflection coefficients of the boundary in the internal reflection case ($n_1>n_2$). (a) TE mode, and (b) TM mode.

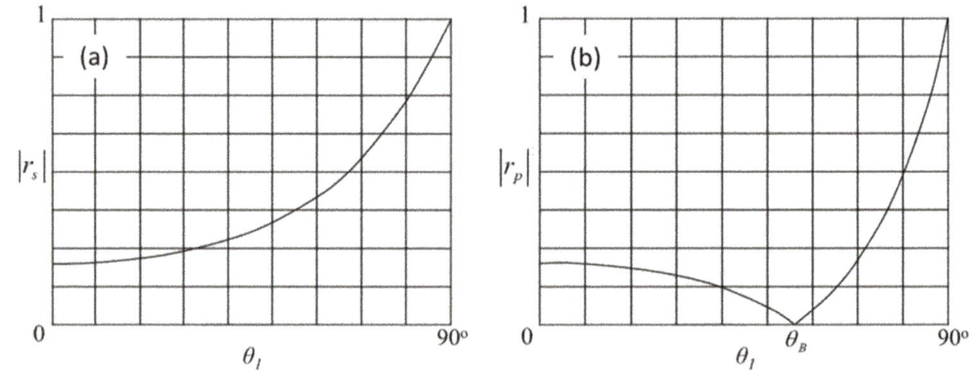

Figure 2.12. Reflection coefficients of the boundary in the external reflection case ($n_1<n_2$). (a) TE mode, and (b) TM mode.

Total reflection

Total reflection from a boundary occurs only in the case of internal reflection when the incident angle satisfies

$$\theta_0 \geqslant \sin^{-1}\left(\frac{n_2}{n_1}\right) \equiv \theta_c, \tag{2.42}$$

where θ_c is called the *critical angle*. Total reflection is an important optical effect, which has been extensively applied. For example, as shown in figure 2.13, a fiber is a solid dielectric cylinder immersed in a surrounding medium of index of refraction lower than that of the fiber so that the light can travel in a fiber through multiple total reflections. According to the critical angle in the boundary between the core of index of refraction n_A and the cladding of index of refraction n_B, we can obtain the maximum semi angle (θ_f) of the vertex of a cone of rays entering the fiber from the end,

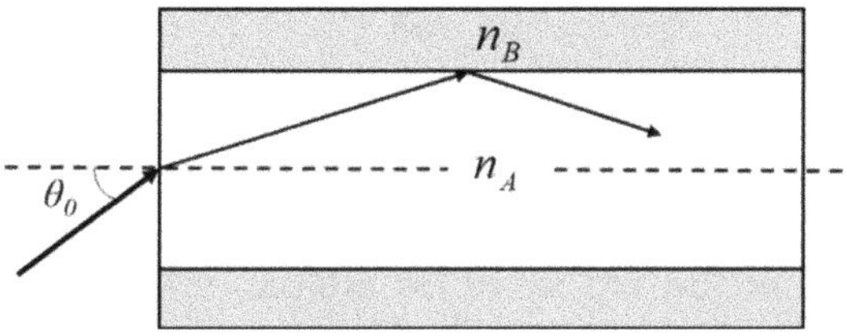

Figure 2.13. A light ray is incident on an optical fiber.

$$\theta_f = \sin^{-1}\sqrt{n_A^2 - n_B^2}, \tag{2.43}$$

and the *numerical aperture* of the fiber

$$NA = \sin\theta_f = \sqrt{n_A^2 - n_B^2}. \tag{2.44}$$

Also, various prisms are applied to change the propagation direction and image orientation with use of total reflection at a boundary instead of coated mirrors.

Total transmission

Total transmission means that no light is reflected from the boundary. We can find that this condition only occurs in the TM case, where the reflection coefficient equals zero when $\theta_0 + \theta_1 = 90^o$. With use of Snell's law, the incidence angle for total transmission is written

$$\theta_0 = \tan^{-1}\left(\frac{n_2}{n_1}\right) \equiv \theta_B, \tag{2.45}$$

where θ_B is called *Brewster angle*. Brewster angle exists in both internal reflection and external reflection, but only in the TM case. Therefore, we can use this effect to purify the polarization state when the light rays are incident on a boundary.

Example 2.1

Given a boundary between the air and a glass ($n = 1.5$), the critical angle for internal reflection is

$$\theta_c = \sin^{-1}\left(\frac{1}{1.5}\right) \approx 41.8°.$$

The Brewster angle of the TM wave for internal reflection is

$$\theta_B = \tan^{-1}\left(\frac{1}{1.5}\right) \approx 33.7°,$$

and for external reflection

$$\theta_B = \tan^{-1}\left(\frac{1.5}{1}\right) \approx 56.3°.$$

Given a boundary between the air and GaAs (n = 3.6), the critical angle for internal reflection is

$$\theta_c = \sin^{-1}\left(\frac{1}{3.6}\right) \approx 16.1°.$$

The Brewster angle of the TM wave for internal reflection is

$$\theta_B = \tan^{-1}\left(\frac{1}{3.6}\right) \approx 15.5°,$$

and for external reflection

$$\theta_B = \tan^{-1}\left(\frac{3.6}{1}\right) \approx 74.5°.$$

2.8 Interaction between light and matter

The discussions in this section, especially those on the diffraction and interference phenomena, are based on the wave theory of light [2, 3]. The propagation of a light wave is governed by the Maxwell equations, and the speed of light is written

$$c = \lambda f. \tag{2.46}$$

At the end of the nineteenth century, the wave theory of light had been challenged by Einstein's observation of the photoelectric effect. The effect addresses that electrons may be emitted from a metal surface when a light ray is incident on it. It was observed that there is a cut-off frequency. The photoelectric effect occurs only when the frequency of the incident ray is higher than the cut-off frequency. To explain this phenomenon, Einstein assumed that light is in a form of energy. He used *photon* to represent the energy of the minimum unit. A photon carries the corresponding energy released from a light source. Each photon has an energy level proportional to the frequency of the light wave, as written

$$E = hf, \tag{2.47}$$

where h is Planck's constant. The higher the frequency of a photon, the more energy it possesses. If the frequency of a photon is reduced below a certain value, no matter how many the photons are in number or how intense the illumination is, there will not be enough energy to eject photoelectrons for the metal. Nowadays, wave–particle duality of light is known by a scientist or an engineer when they process a study of light waves or photons. The diffraction and interference of light are better explained by the wave nature, but the interaction between light and matter is better interpreted by using the concept of photons.

Example 2.2

Photon energy is often measured in terms of electron-volts [eV]. If the wavelength of a light wave is 1 μm, calculate the photon energy in terms of the electron-volt.

From equation (2.47), we have

$$E = hf = \frac{hc}{\lambda} = \frac{6.63 \times 10^{-34}\,(\mathrm{J \cdot s}) \times 3 \times 10^{8}(\mathrm{m\ s^{-1}})}{1 \times 10^{-6}\,(\mathrm{m})} = 2 \times 10^{-19}\ \mathrm{J}.$$

One electron-volt is the energy required to raise the potential of an electron by one volt. Since the charge of an electron is 1.6×10^{-19} coulomb [C], 1 eV corresponds to

$$1.6 \times 10^{-19}\,(\mathrm{C}) \times 1\,(\mathrm{V}) = 1.6 \times 10^{-19}\ \mathrm{J},$$

and the photon energy can be expressed as

$$E = \frac{2 \times 10^{-19}\,(\mathrm{J})}{1.6 \times 10^{-19}\,(\mathrm{J})} = 1.24\ \mathrm{eV}.$$

To estimate the photon energy for a given wavelength λ, the following equation can be used:

$$E = \frac{1.24}{\lambda},$$

where the energy is measured in electron-volts and the unit of wavelength is micrometers.

2.9 Basic principle of geometrical optics

No matter whether a ray propagates in a free space or from a medium to another medium, the propagation direction is well defined. The unique propagation direction can be determined by Fermat' s principle, where a ray propagating from one point to another point follows the path for the shortest traveling time, or equivalently the shortest optical path length or shortest accumulated phase. Therefore, when a ray propagates in a free space, the path from a point to another point is a straight line connecting the two points. We can further determine the light path from a medium to another one through Fermat's principle [3]. As shown in figure 2.14, the ray

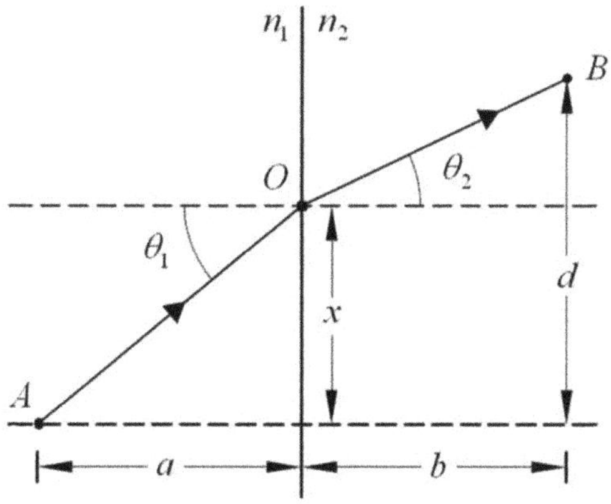

Figure 2.14. Light propagation according to Fermat's principle.

propagates from points A to another point B, where the refractive index around points A and B is n_1 and n_2, respectively. Assume that the light ray passes through a point O located at the interface, then the propagation time from point A to point B is written

$$t = \frac{AO}{c/n_1} + \frac{OB}{c/n_2}, \tag{2.48}$$

where

$$AO = \sqrt{a^2 + x^2}$$

$$OB = \sqrt{b^2 + (d - x)^2}$$

According to the principle, the position of O will be determined with $\frac{dt}{dx} = 0$. Therefore, we obtain

$$\frac{n_1 x}{c\sqrt{a^2 + x^2}} - \frac{n_2(d - x)}{c\sqrt{b^2 + (d - x)^2}} = 0. \tag{2.49}$$

We can further assume that the incident angles with respect to the surface normal are θ_1 and θ_2, respectively. Then we obtain the equation for the position of O,

$$n_1 \sin\theta_1 = n_2 \sin\theta_2. \tag{2.50}$$

Equation (2.50) is called Snell's Law for refraction. It should be noted that Snell's law concerns the propagation direction of the wavefront normal, i.e., the wave vector. In an isotropic medium, the direction of the wave vector coincides with the direction of the Poynting vector. In contrast, the direction between the wave vector and the Poynting vector could be different in an anisotropic medium, where Snell's law cannot be used to describe the propagation direction of the ray, which represents the Poynting vector. Similarly, Fermat's principle can describe the reflected ray by a mirror. As shown in figure 2.15, A and B are located at the same medium, point O on the surface is responsible to the reflection of the light ray from point A to point B. Through the same calculation, we obtain the equation for determining the position of point O

$$\theta_2 = \theta_1. \tag{2.51}$$

Equation (2.51) is called the principle of reflection. It should be noted that both the principles of the fraction and reflection can be derived from Maxwell's equations. Another important result is that those points A, O, B are located on the same plane, which is called the incident plane. The three propagation properties described above, rectilinear propagation, refraction and reflection are the basics of geometrical optics.

2.10 Mirrors and lenses

In the following, we will introduce some basic optical components [3]. First we need to introduce the optical axis, which serves as the reference line in an optical system, and it is a common axis of rotation-symmetry for an optical system consisting of a series of refracting and/or reflection surfaces.

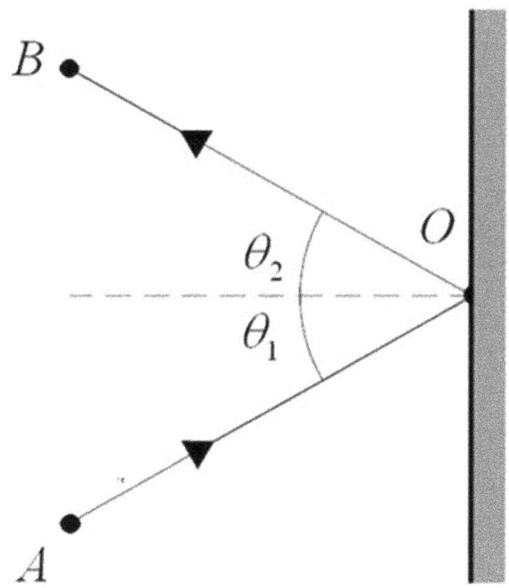

Figure 2.15. Principle of reflection.

Planar mirrors

A planar mirror is the simplest and most important optical component, which is used to change the propagation direction of a light wave without changing the wavefront. As shown in figure 2.16, one can see the virtual image of the object on the same side through a planar mirror.

Elliptical mirrors

An elliptical mirror shown in figure 2.17 has two focal points. Light emerging from one focal point will focus at the other focal point through reflection of the elliptical surface.

Hyperbolic mirrors

A hyperbolic mirror shown in figure 2.18 has a pair of focal points. A divergent spherical wave emerging from one focal point will look like emerging from the other through reflection. In contrast, a converging wave toward one of the focal points will focus at the other through reflection.

Parabolic mirrors

A parabolic mirror shown in figure 2.19 has a well-defined focus point. Collimating rays are focused at the focus point through the reflection from a hyperbolic mirror, and reversed rays emitted from the focus point become collimated through the reflection.

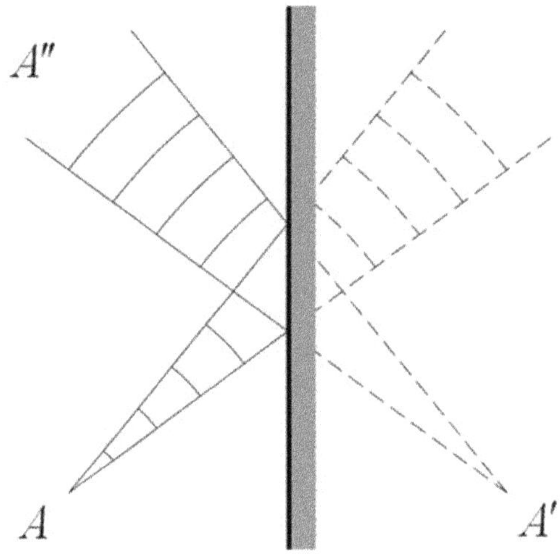

Figure 2.16. A planar mirror.

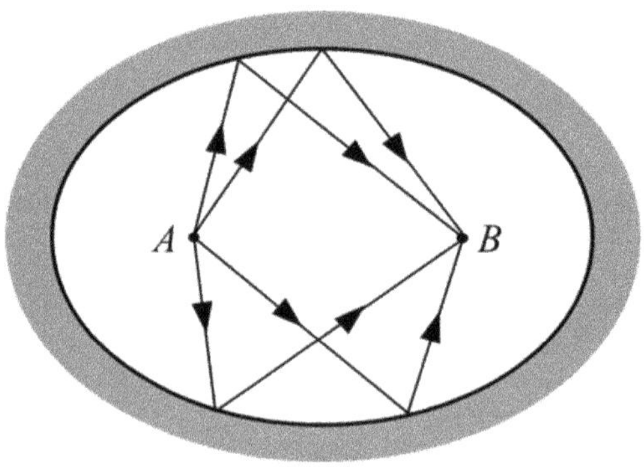

Figure 2.17. An elliptical mirror.

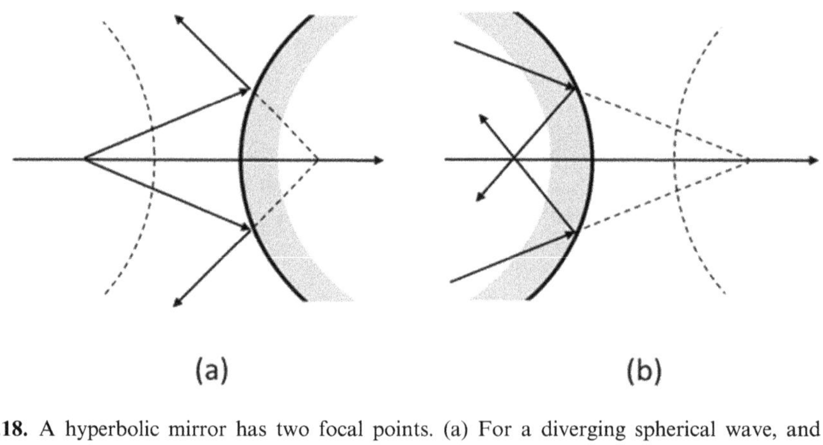

Figure 2.18. A hyperbolic mirror has two focal points. (a) For a diverging spherical wave, and (b) for a converging spherical wave.

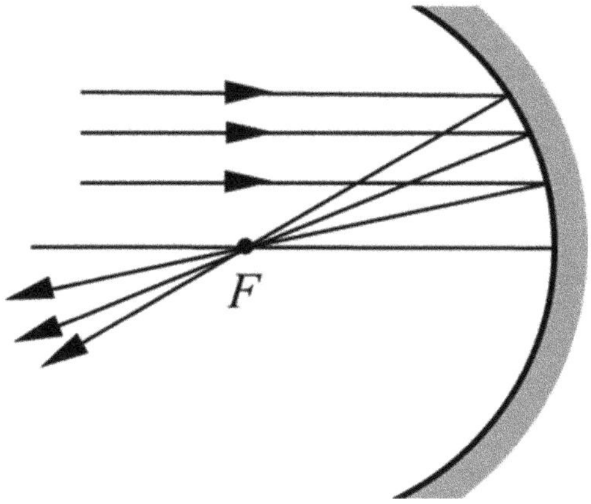

Figure 2.19. A parabolic mirror.

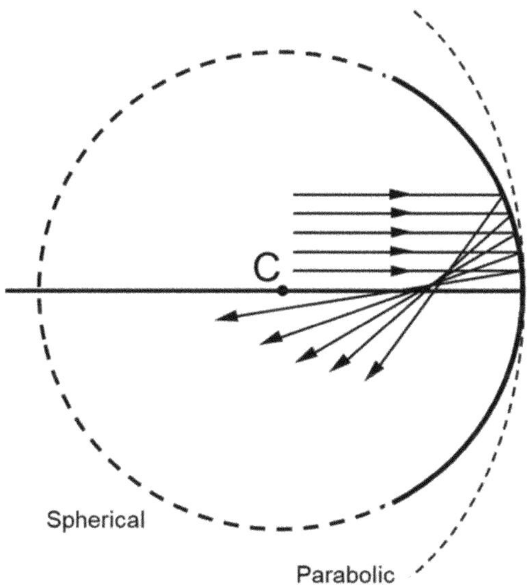

Figure 2.20. A spherical mirror.

Spherical mirrors

Spherical mirrors, as shown in figure 2.20, are similar to parabolic mirrors except the focus point is not well defined. This phenomenon becomes obvious when the rays are incident on the further side of the spherical mirror. When the rays are incident on the part near the center of the spherical mirror, the reflected rays intersect around a point, which is called the paraxial focus point of a spherical mirror. Therefore, a spherical mirror could nearly have a focus point under the paraxial condition, where

$$\sin \theta \approx \theta \tag{2.52}$$

holds for the incident angle shown in figure 2.21.

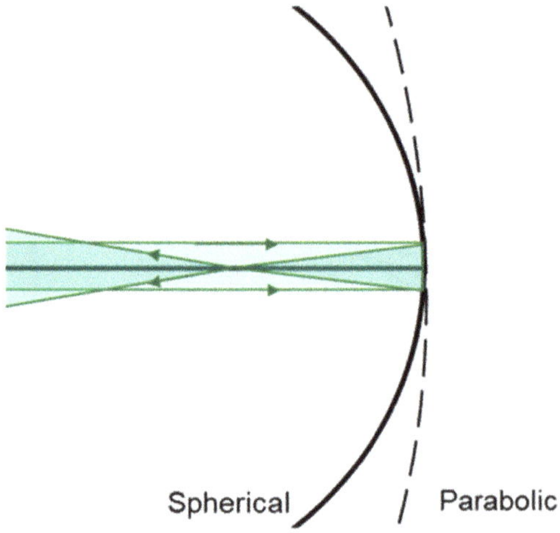

Figure 2.21. The paraxial condition of spherical mirror.

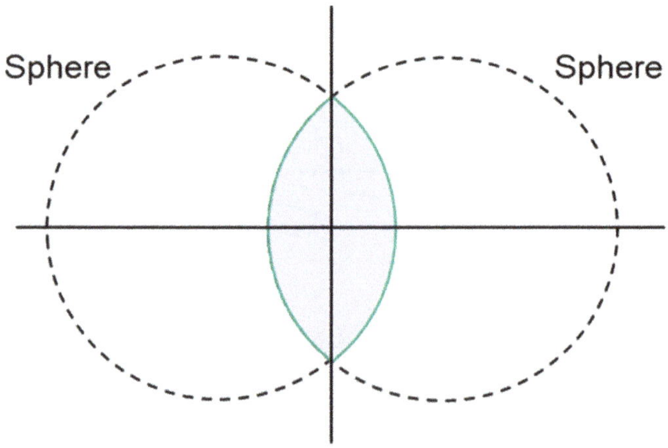

Figure 2.22. A spherical lens with two spherical surfaces.

Spherical lenses

Spherical lenses are the most common optical elements with one or two spherical surfaces, shown in figure 2.22. The function is similar to spherical mirrors except the directions of incident light change according to refraction instead of reflection. Typically, a spherical lens has a focus point under the paraxial condition. Beyond the paraxial condition, a spherical lens can never have a well-defined focal point, as shown in figure 2.23. Therefore, collimating rays cannot be focused at a point, and this condition causes blur in image formation. Accordingly, an aspherical lens with aspherical surfaces is developed to obtain a better image quality.

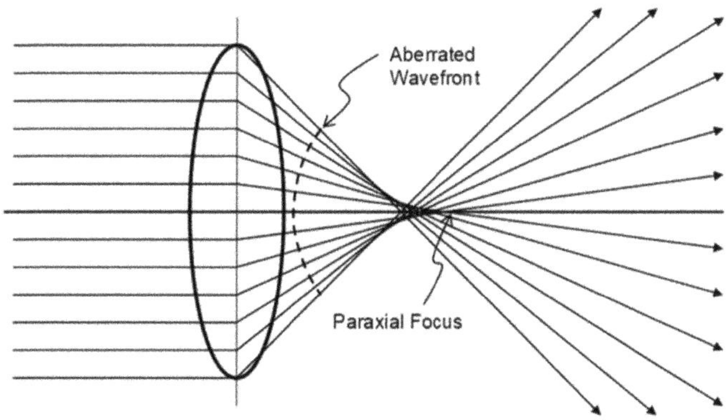

Figure 2.23. A lens cannot direct all the parallel incident rays into a focus when the lens has aberrations.

2.11 Prisms

A prism is a usual volume of optical medium, e.g., glass, containing several planar surfaces [3]. Due to the difference in geometry, prisms may change the light propagation direction through multiple refractions or internal total reflection. The most interesting thing is that the image orientation may be changed.

Minimum deviation

The propagation angle of the light incident on the left side of the prism indicated in figure 2.24 changes through two refractions on the entrance and exit surface. The angle deviation δ of the entrance and the exit angle can be obtain

$$\delta = \theta_1 + \theta_4 - \alpha, \tag{2.53}$$

where α is the apex angle of the prism, and

$$\alpha = \theta_2 + \theta_3. \tag{2.54}$$

The apex angle is fixed, so we can obtain

$$d\theta_2 = -d\theta_3. \tag{2.55}$$

For a minimum deviation,

$$\frac{d\delta}{d\theta_1} = 1 + \frac{d\theta_4}{d\theta_1} = 0, \tag{2.56}$$

and then

$$d\theta_4 = -d\theta_1. \tag{2.57}$$

By applying Snell's law to the refraction of the entrance and the exit surfaces, we obtain

$$\sin\theta_1 = n\sin\theta_2 \tag{2.58}$$

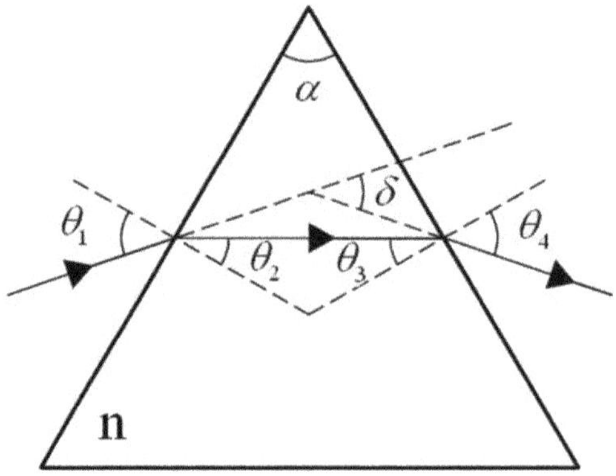

Figure 2.24. A prism for measuring minimum angular deviation and the refractive index.

and

$$n \sin \theta_3 = \sin \theta_4. \tag{2.59}$$

With equations (2.55) and (2.57), the differentiation of equations (2.58) and (2.59) can be written

$$\frac{\cos \theta_1}{\cos \theta_4} = \frac{\cos \theta_2}{\cos \theta_3}. \tag{2.60}$$

Applying equations (2.58) and (2.59), equation (2.60) can be rewritten

$$\frac{1 - \sin^2 \theta_1}{1 - \sin^2 \theta_4} = \frac{n^2 - \sin^2 \theta_1}{n^2 - \sin^2 \theta_4}. \tag{2.61}$$

Equation (2.61) holds for different refractive indices, so the solution is

$$\theta_4 = \theta_1,$$

$$\theta_3 = \theta_2. \tag{2.62}$$

Therefore, for the minimum deviation the incident angle is

$$\theta_1 = \frac{\delta_{\min} + \alpha}{2}. \tag{2.63}$$

Then the refractive index can be expressed as a function of the minimum deviation

$$n = \frac{\sin \dfrac{\delta_{\min} + \alpha}{2}}{\sin \dfrac{\alpha}{2}}. \tag{2.64}$$

Reflecting prisms

In the following, we will introduce several prisms used practically in optical devices. The total internal reflection is the main mechanism of the reflecting surface in a prism and one of the main purposes for the use of a prism.

The right-angle prism

As shown in figure 2.25, the right-angle prism contains a reflecting surface at 45° with respect to two orthogonal surfaces. The ray in normal incidence on the entrance face will be reflected by the reflecting surface with total internal reflection and then is normally incident on the exit face. It should be noted that a right-hand oriented image will be changed to a left-hand one.

The Dove prism

The Dove prism is used to change the orientation of the image but the propagation direction is kept unchanged, as shown in figure 2.26.

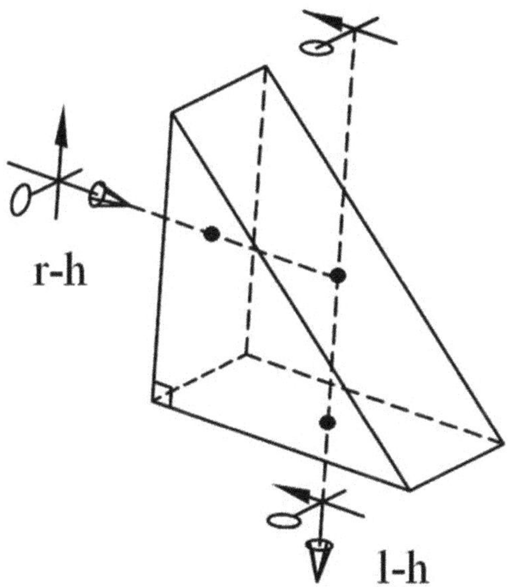

Figure 2.25. The right-angle prism.

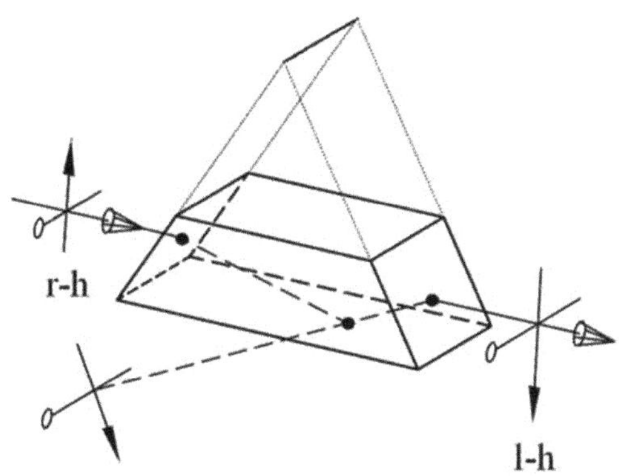

Figure 2.26. The Dove prism.

The Porro prism

The Porro prism is the same as the right-angle prism. However, the entrance and exit faces are the same, as shown in figure 2.27. The two reflecting surfaces are orthogonal, just like a corner cube, so the reflected light parallels the incident light, and the image orientation does not change. The Porro prism may act as a reversed reflector if the dimensions are small, where the reflected light almost reverses the path of the incident light.

The penta prism

The penta prism, as indicated in figure 2.28, can reflect light in 90° but keep the orientation of the image unchanged. It is noted that the optical path length in a penta prism is extended in comparison with a right-angle prism of a mirror.

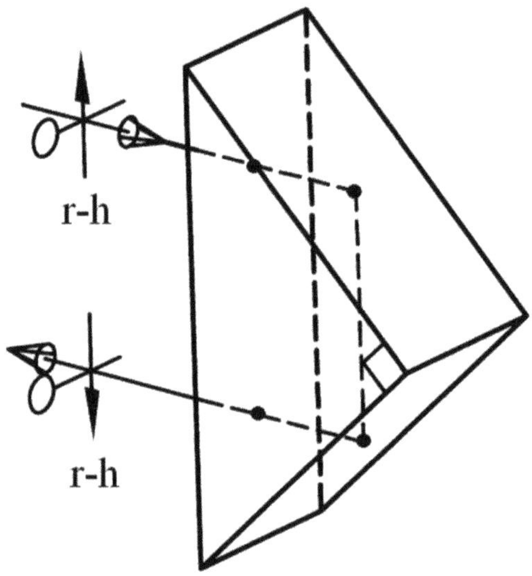

Figure 2.27. The Porro prism.

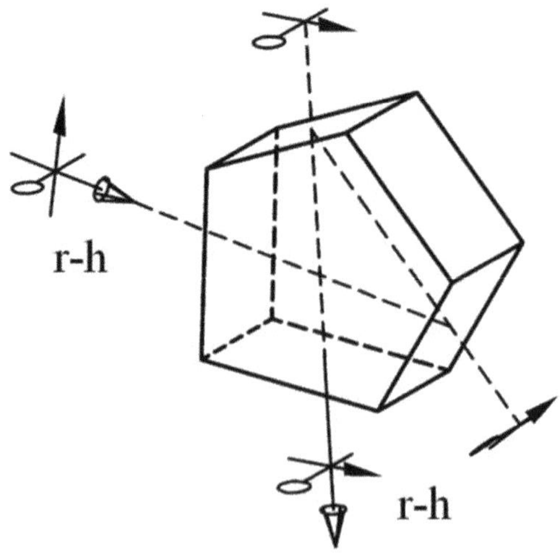

Figure 2.28. The penta prism.

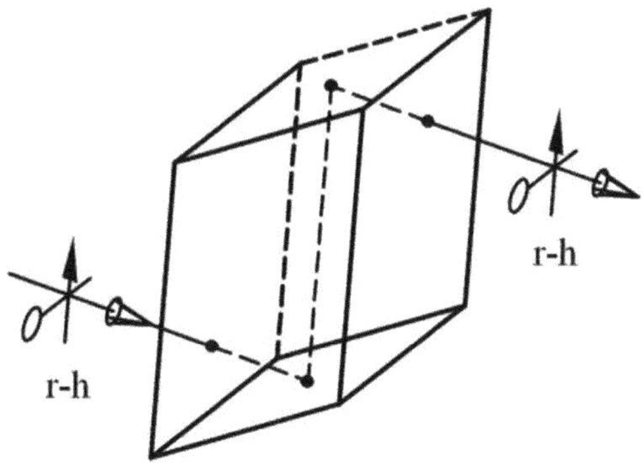

Figure 2.29. Rhomboid prism.

The rhomboid prism

The rhomboid prism contains two parallel reflecting surfaces, as indicated in figure 2.29, and can be used to laterally displace the incident light and keep the orientation of the carried image.

2.12 Gaussian optics

Gaussian optics describes an approximation of a theory for the light propagating at an angle satisfying the paraxial condition [3, 6, 7]. Under the condition, both a spherical mirror and a spherical lens are said to have a well-defined focus point. In the following, we will introduce imaging by optical elements in the scope of Gaussian optics. For convenience, we list the rules of the sign convention that we will encounter in the following discussions of Gaussian optics.

1. Rays are incident on a system from left to right.
2. Distances to the right of and above a reference point are positive; otherwise, the distances are negative.
3. The radius of curvature of a surface is regarded as the distance of its center of curvature from its vertex. The radius of curvature is positive (negative) when the center of curvature lies to the right (left) of the vertex.
4. The acute angle of a ray from the optical axis or from the surface normal is positive (negative) if it is counterclockwise (clockwise).
5. The refractive index and the distance between two adjacent surfaces are given a negative sign when a ray is propagated from right to left.

Spherical refracting surface

Shown in figure 2.30 is a spherical refracting surface of a radius of curvature R separating media of refractive indices n_1 and n_2. Assume that the object point is at P_1 and the image point is at P_2. A ray from the object is incident on a point Q at the surface and then is refracted to the image. The line CQ connecting Q and the center of the curvature C is the normal axis of the incident point at Q.

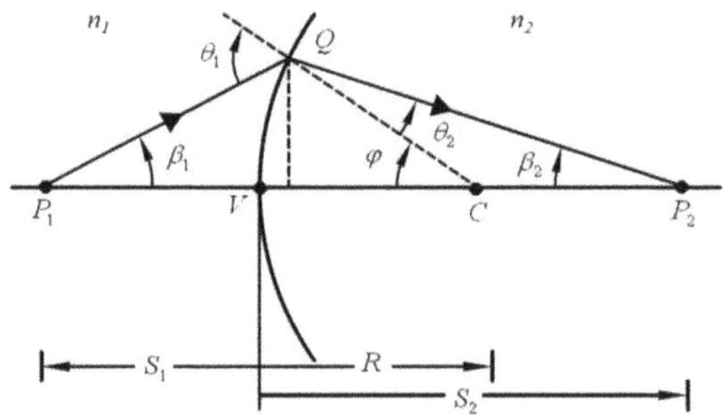

Figure 2.30. A spherical refracting surface.

In the Gaussian approximation of Snell's law, the relation between the incident angle and the refraction angle is written

$$n_1\theta_1 = n_2\theta_2. \tag{2.65}$$

According to the geometry, we note that

$$\theta_1 = \beta_1 - \varphi \tag{2.66}$$

and

$$\theta_2 = \beta_2 - \varphi. \tag{2.67}$$

Under the paraxial approximation, we obtain

$$\beta_1 = -\frac{x}{S_1} \tag{2.68}$$

$$\beta_2 = -\frac{x}{S_2} \tag{2.69}$$

and

$$\varphi = -\frac{x}{R}, \tag{2.70}$$

where S_1, S_2, and R are the object distance, image distance and the radius of curvature of the surface, respectively. We obtain the imaging equation from equations (2.65)–(2.70)

$$\frac{n_2}{S_2} - \frac{n_1}{S_1} = \frac{n_2 - n_1}{R}. \tag{2.71}$$

Equation (2.71) is called the Gaussian *imaging equation.* The point object P_1 and its corresponding Gaussian image point P_2 are called conjugate points. When the object lies at infinity, i.e., $S_1 \to -\infty$, the corresponding imaging point is located at the focal point of the spherical refracting surface. Then the focal length in the image space can be obtained with equation (2.71)

$$f_2 = \frac{n_2}{n_2 - n_1}R, \tag{2.72}$$

and in the object space

$$f_1 = -\frac{n_1}{n_2 - n_1}R. \tag{2.73}$$

From the above equations, we find that focal length varies when the refractive index around the focal point changes. Therefore, to obtain a unique value of focal length, we define the *effective focal length*

$$f_e = \frac{f_2}{n_2} = -\frac{f_1}{n_1} = \frac{1}{K}, \tag{2.74}$$

where K is called *refracting power.* A refracting surface is called a positive or converging surface when the focal power and the effective length are positive; in contrast, a refracting surface is called a negative or divergent surface when it is negative.

To image an extended object, the magnification of the image is an important factor. As shown in figure 2.31, we find that the transverse magnification and the angular magnification can be respectively written

$$M_t = \frac{h_2}{h_1} = \frac{n_1 S_2}{n_2 S_1}, \tag{2.75}$$

$$M_\beta = \frac{\beta_2}{\beta_1} = \frac{S_1}{S_2}. \tag{2.76}$$

Equations (2.75) and (2.76) mean that a large transverse magnification of the image can be obtained only with a small angular magnification of the rays. These two equations can also be written

$$n_2 h_2 \beta_2 = n_1 h_1 \beta_1. \tag{2.77}$$

Equation (2.77) shows that the product of the refractive index, image height and ray angle do not change upon refraction. This is called Lagrange invariant.

In cases of a small change ΔS_1 is introduced in the object distance, a corresponding change ΔS_2 will result in the image distance. The ratio $\frac{\Delta S_2}{\Delta S_1}$ is called the longitudinal magnification

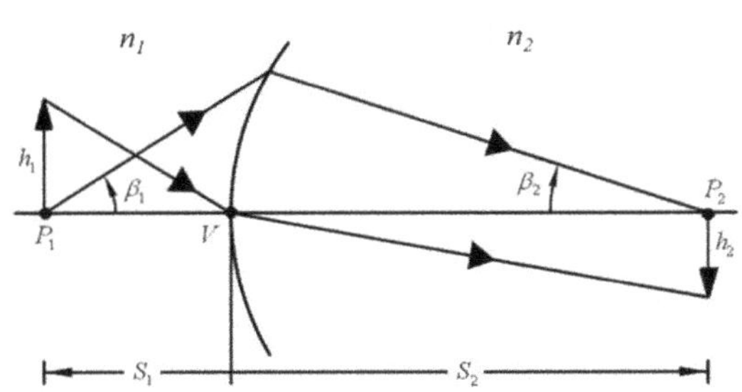

Figure 2.31. Transverse magnification.

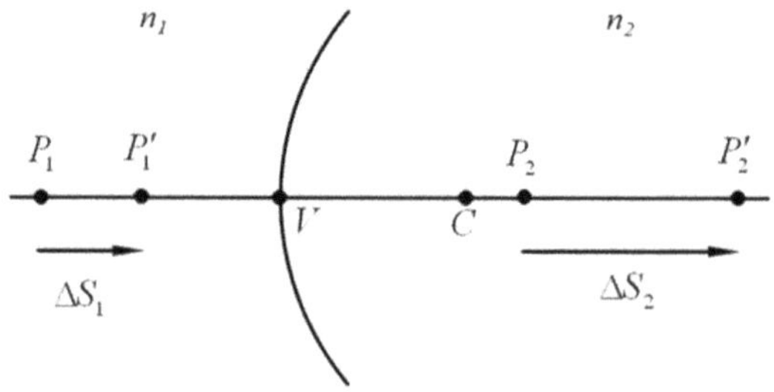

Figure 2.32. Longitudinal magnification.

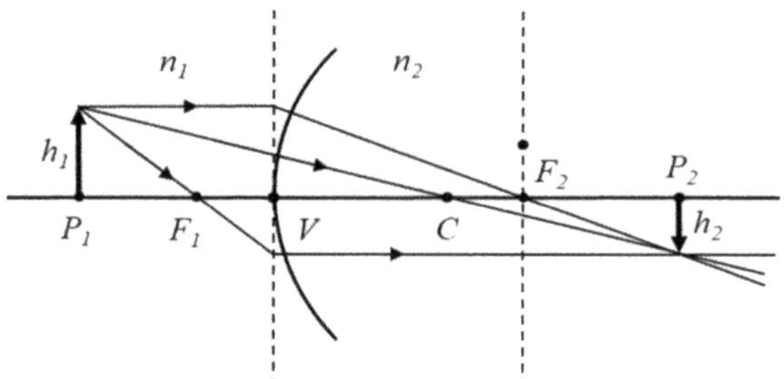

Figure 2.33. The three rays for locating the image.

$$M_L = \frac{\Delta S_2}{\Delta S_1} = \frac{M_t}{M_\beta} = \frac{n_1}{n_2} M_t^2. \tag{2.78}$$

Equation (2.78) shows that the image moves along the same direction as the object does, as described in figure 2.32. If a lens is used to image a 3-D object, equation (2.78) also indicates that a distortion image will be formed due to different magnification in the longitudinal direction with respect to the other direction.

A simple way called graphical imaging can be used to locate an image easily. As illustrated in figure 2.33, three rays emitted from an object point can be traced without precise calculation. The first ray incident parallel to the optical axis passes through the back focal point F_2 after refraction. The second ray directed to the center of the curvature C does not change the propagation direction. The third ray passing through the front focal point F_1 becomes parallel to the optical axis. These three rays intersect at the corresponding image point. Generally, two rays are enough to locate the image. There is another graphical method to locate the image of an on-axis object point based on the principle that all parallel rays intersect at a point on the back focal plane. As illustrated in figure 2.34, we can use a ray, which is parallel to the object ray, directing toward the center of curvature to locate the interesting point I on the back focal plane. Thus the object ray passes the point I through refraction and then intersects the optical axis at the corresponding image point.

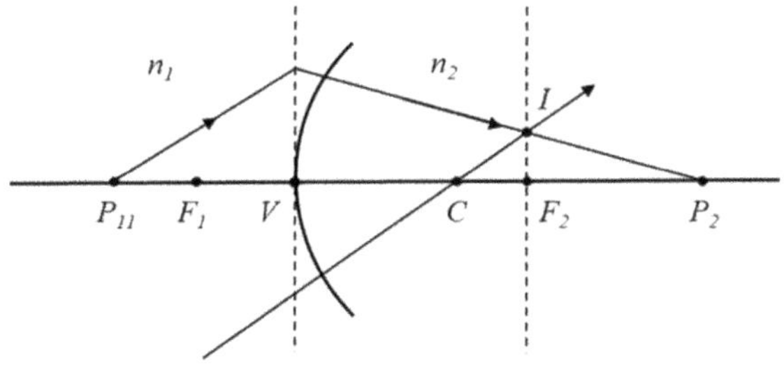

Figure 2.34. The fourth ray, which is parallel to the incident ray, is used to indicate the path of the refraction ray of the lens.

2.13 Thin lens

A thin lens is regarded as two refraction surfaces with a separating distance which can be negligible [6, 7]. Consider a thin lens made with a material with a refractive index n_L, and the refractive index of the material around the lens is n_A, as shown in figure 2.35. Let the radii of the curvature of the surfaces be R_1 and R_2. According to equation (2.71), we can obtain the imaging equation for the first surface

$$\frac{n_L}{S_{12}} - \frac{n_A}{S_{11}} = \frac{n_L - n_A}{R_1}, \tag{2.79}$$

and for the second surface

$$\frac{n_A}{S_{22}} - \frac{n_L}{S_{21}} = \frac{n_A - n_L}{R_2}, \tag{2.80}$$

where S_{11} and S_{12} are the object distance and image distance of the first surface, respectively; S_{21} and S_{22} are the object distance and image distance of the second, surface respectively. We note $S_{21} = S_{12}$ for a thin lens. Combining equations (2.79) and (2.80), we obtain

$$\frac{1}{S_2} - \frac{1}{S_1} = \frac{n_L - n_A}{n_A}\left(\frac{1}{R_1} - \frac{1}{R_2}\right). \tag{2.81}$$

To obtain the focal length, we let $S_1 \to \infty$ and image space focal length is obtained

$$\frac{1}{f_2} = \frac{n_L - n_A}{n_A}\left(\frac{1}{R_1} - \frac{1}{R_2}\right). \tag{2.82}$$

If the lens is put in air, i.e., $n_A = 1$, then the focal length in air can be obtained

$$\frac{1}{f_e} = (n_L - 1)\left(\frac{1}{R_1} - \frac{1}{R_2}\right) \equiv K. \tag{2.83}$$

Equation (2.83) is called the lens maker's formula. Combining equations (2.81) and (2.83), the imaging equation of a thin lens can be rewritten

$$\frac{1}{S_2} - \frac{1}{S_1} = \frac{1}{f_e}. \tag{2.84}$$

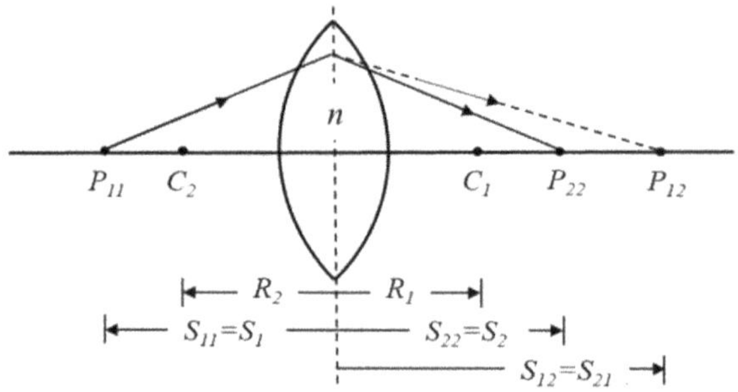

Figure 2.35. Imaging of a thin lens.

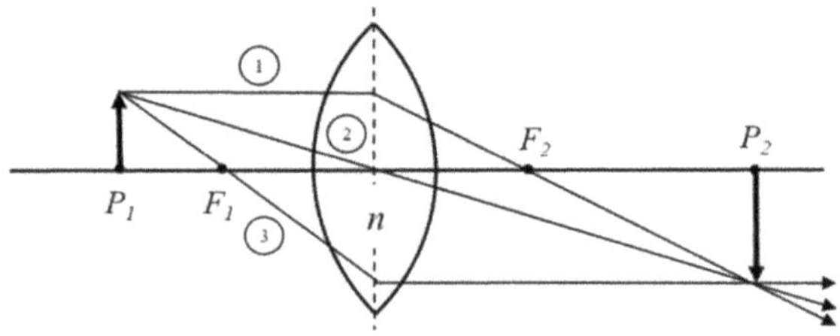

Figure 2.36. The three rays for imaging in the case of a lens.

Graphical imaging is also useful to determine the image point in a thin lens. As illustrated in figure 2.36, the principles of the first ray and the third ray are the same as that in a refracting surface except that the second ray, which directs toward the center of the thin lens does not change the propagation direction.

2.14 Thick lens

In contrast to a thin lens, the separating distance between two refracting surfaces of a thick lens cannot be neglected [6]. Thus, we have to calculate the change in the position of light for the transmission between two surfaces. In a thick lens as well as a lens system, there are six cardinal points, including two principal points, two focal points and two nodal points used to characterize the imaging property. Figure 2.37 indicates that a light directed to a nodal point will emerge from the other nodal point so that the angular magnification is 1. The locations of the principal point can be obtained as illustrated in figure 2.38. A light propagating parallel to the optical axis incident on a lens passes through the back focal point when it emerges from the lens. The extension of the incident and the emergent rays intersect at Q. A plane containing Q and normal to the optical axis is called the principal plane, where the intersect point is the principal point. It should be noted that the principal points are the reference points in both object and image space instead of apex points of the lens surfaces. The object distance is measured from the front principal point to the object and the image distance is from the back principal point to the object. Similarly, the object space focal length is from the front principal point to the object

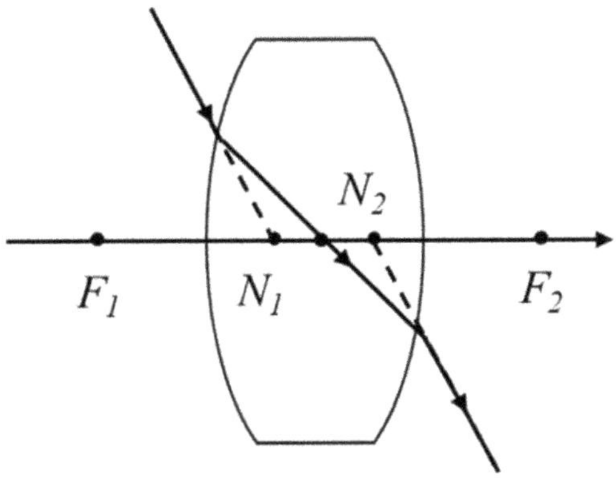

Figure 2.37. Parallel transmitting rays in a thick lens.

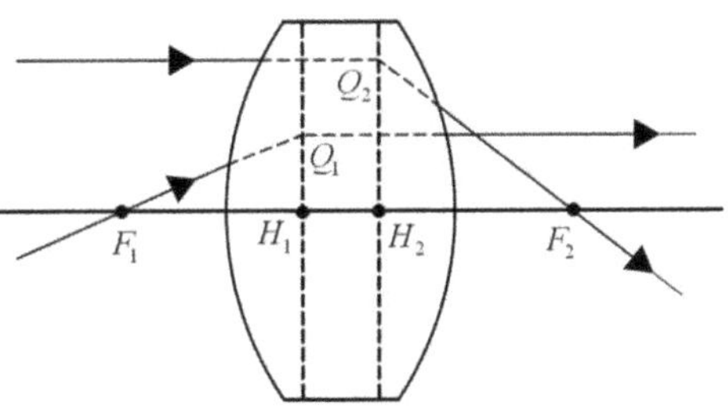

Figure 2.38. Two principal planes in a thick lens.

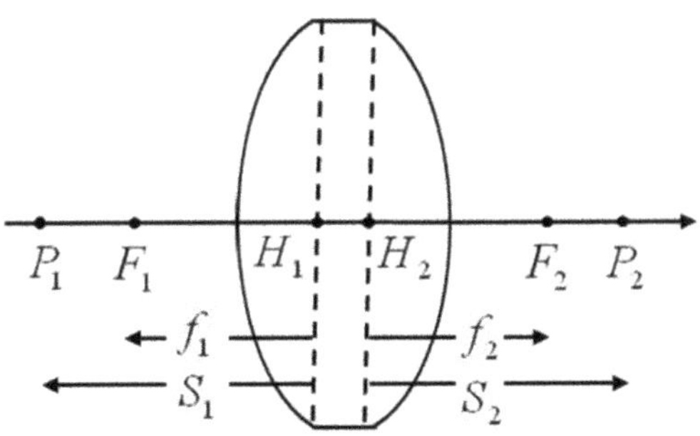

Figure 2.39. The locations and distance definition of a thick lens.

space focal point and the image space focal length is from the back principal point to the image space focal point, as shown in figure 2.39.

The locations of the principal points in various lenses is shown in figure 2.40. Generally, the locations of the principal points and the focal points are sufficient to describe Gaussian imaging of the system.

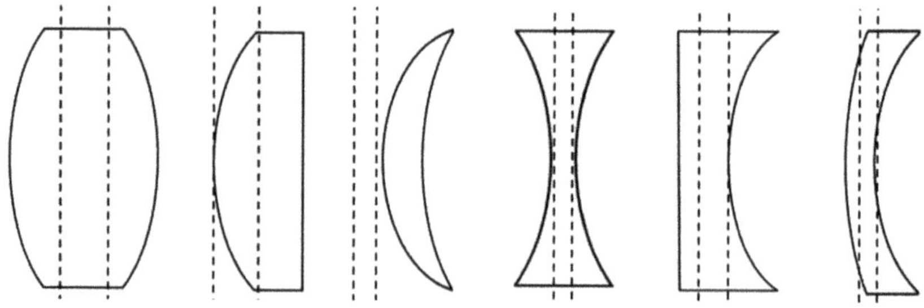

Figure 2.40. (a)–(c) Convex lens, (d)–(f) concave lens.

2.15 Spherical mirror

A spherical mirror reflects incident rays and images an object through reflection. As indicated in figure 2.41, the light emerging from P_1 arrives at P_2 from the reflection at Q. According to reflection law,

$$\theta_2 = -\theta_1. \tag{2.85}$$

From triangles P_1QC and QCP_2, we have

$$\phi = \beta_1 - \theta_1, \tag{2.86}$$

and

$$\beta_2 = \phi + \theta_2. \tag{2.87}$$

Under the paraxial condition,

$$\beta_1 = -\frac{x}{S_1}, \tag{2.88}$$

$$\beta_2 = -\frac{x}{S_2}, \tag{2.89}$$

and

$$\phi = -\frac{x}{R}. \tag{2.90}$$

Substituting equations (2.86) and (2.87) to equation (2.85), we obtain

$$\frac{1}{S_2} + \frac{1}{S_1} = \frac{2}{R}. \tag{2.91}$$

When the object is located at infinity, the focal length of the mirror is

$$f = \frac{R}{2}. \tag{2.92}$$

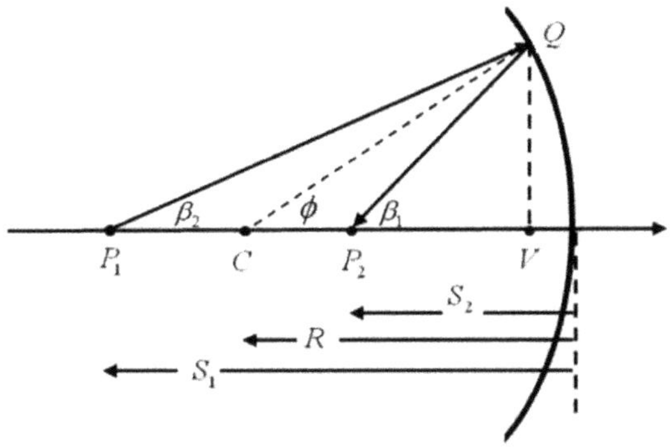

Figure 2.41. Imaging of a spherical mirror.

The focal length is half of the radius of curvature, and this means that the focal point lies at the middle between V and C.

2.16 Paraxial ray tracing

An optical system may contain multiple refracting or reflecting surfaces. The scheme to trace both the location and the propagation direction of a ray is called ray tracing. Ray tracing is one of the most effective schemes to figure out the light path and obtain the image quality of an optical imaging system. Paraxial ray tracing is used for tracing rays under paraxial conditions [3, 6, 7]. If the propagation rays are beyond the paraxial approximation, a true ray tracing should replace the paraxial one. In the following, we will introduce paraxial ray tracing based on Gaussian optics.

In ray tracing, we are concerned with both the height and the propagation direction of a light ray at a certain location. Thus we define a *ray vector*

$$u = \begin{pmatrix} x \\ n\beta \end{pmatrix}, \tag{2.93}$$

where x is the height of the ray and β is the angle with respect to the optical axis. In the case that a ray propagates from a point to another point, the ray vector can be expressed

$$\begin{pmatrix} x_2 \\ n_2\beta_2 \end{pmatrix} = T_{12}\begin{pmatrix} x_1 \\ n_1\beta_1 \end{pmatrix}, \tag{2.94}$$

where T_{12} is called the transfer matrix, and is written

$$T_{12} = \begin{pmatrix} 1 & \dfrac{t_{12}}{n} \\ 0 & 1 \end{pmatrix}. \tag{2.95}$$

In the case when a ray is refracted by a spherical surface, the ray vector can be expressed

$$\begin{pmatrix} x_2 \\ n_2\beta_2 \end{pmatrix} = R_{12}\begin{pmatrix} x_1 \\ n_1\beta_1 \end{pmatrix}, \tag{2.96}$$

where R_{12} is called the refracting matrix, and is written

$$R_{12} = \begin{pmatrix} 1 & 0 \\ \dfrac{n_1 - n_2}{R} & 1 \end{pmatrix} = \begin{pmatrix} 1 & 0 \\ -K & 0 \end{pmatrix}. \tag{2.97}$$

In the case that a ray is reflected by a spherical mirror, the ray vector is written

$$\begin{pmatrix} x_2 \\ n_1\beta_2 \end{pmatrix} = M_{12}\begin{pmatrix} x_1 \\ n_1\beta_1 \end{pmatrix}, \tag{2.98}$$

where M_{12} is called the reflecting matrix, and is written

$$M_{12} = \begin{pmatrix} 1 & 0 \\ -\dfrac{2}{R} & 1 \end{pmatrix}. \tag{2.99}$$

Example 2.3

Find the locations of the principal points, focal points and the distances between them of a thick lens shown in figure 2.42 with use of the matrix approach for paraxial ray tracing.

Let the ray vector of the incident ray be

$$u_1 = \begin{pmatrix} x_1 \\ 0 \end{pmatrix}.$$

Then the ray vector at P_4 is

$$u_4 = \begin{pmatrix} x_4 \\ \beta_4 \end{pmatrix} = T_{34}R_3T_{23}R_2T_{12}u_1$$

$$= \begin{pmatrix} 1 & t_3 \\ 0 & 1 \end{pmatrix}\begin{pmatrix} 1 & 0 \\ \dfrac{n-1}{R_3} & 1 \end{pmatrix}\begin{pmatrix} 1 & \dfrac{t_2}{n} \\ 0 & 1 \end{pmatrix}\begin{pmatrix} 1 & 0 \\ \dfrac{1-n}{R_2} & 1 \end{pmatrix}\begin{pmatrix} 1 & t_1 \\ 0 & 1 \end{pmatrix}\begin{pmatrix} x_1 \\ 0 \end{pmatrix}$$

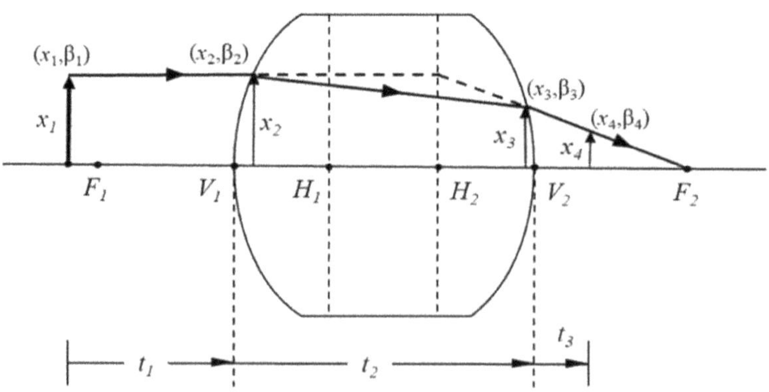

Figure 2.42. Ray tracing in example 2.3.

Because β_4 is the ratio of x_1 and the effective focal length, we can obtain the image space focal length

$$\frac{1}{f_2} = (n-1)\left(\frac{1}{R_2} - \frac{1}{R_3}\right) + \frac{t_2(n-1)^2}{R_2R_3}. \tag{2.100}$$

The back focal length V_2F_2 can be obtained by the ratio of x_3 and β_4

$$V_2F_2 = f_2\left(1 - \frac{n-1}{nR_2}t_2\right). \tag{2.101}$$

Therefore, H_2V_2 is obtained

$$V_2H_2 = -(f_2 - V_2F_2) = -\frac{n-1}{nR_2}t_2 f_2. \tag{2.102}$$

With use of the similar calculation, we can obtain the front focal length

$$V_1F_1 = f_2\left(1 + \frac{n-1}{nR_2}t_2\right), \tag{2.103}$$

and

$$V_1H_1 = -\frac{n-1}{nR_3}t_2 f_2. \tag{2.104}$$

Example 2.4

Find the image location of a glass hemisphere shown in figure 2.43 for an object is at a distance $4\frac{2}{3}$ cm in front of the planar surface of the lens. The radius of the curvature of the second surface is 2 cm, and the refractive index of the glass is 1.5.

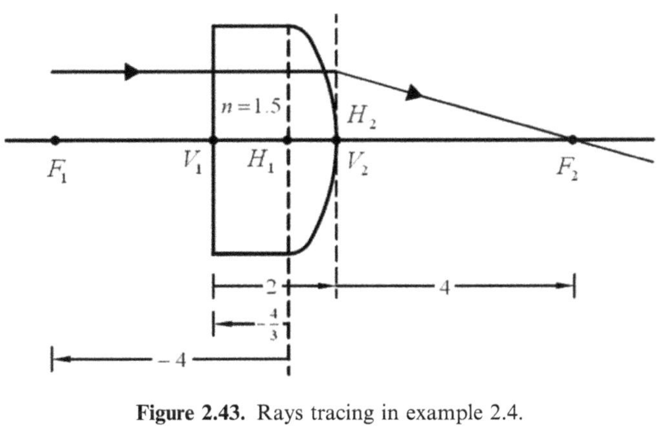

Figure 2.43. Rays tracing in example 2.4.

From equation (2.100)

$$\frac{1}{f_2} = (1.5 - 1)\left(\frac{1}{\infty} - \frac{1}{-2}\right) + \frac{2(1.5 - 1)^2}{(1.5)(\infty)(-2)} = \frac{1}{4} = -\frac{1}{f_1}.$$

Therefore, the effective focal length is 4 cm.

From equation (2.102)

$$V_2H_2 = 0.$$

From equation (2.104)

$$V_1H_1 = -\frac{(1.5 - 1)}{(1.5)(-2)}(2)(4) = \frac{4}{3}.$$

The locations of the principal points and focal points are shown in figure 2.42. Thus the object distance

$$S_1 = H_1P_1 = V_1P_1 + V_1H_1 = 6.$$

Applying equation (2.86), we obtain the corresponding image distance

$$S_2 = H_2P_2 = 12.$$

Therefore, the image is at the other side of the lens at a distance 12 cm from the back apex point which is at the same position of the back principal point.

References

[1] Born M and Wolf E 2005 *Principles of Optics* (Cambridge: Cambridge University Press)

[2] Saleh B E A and Teich M C 1991 *Fundamentals of Photonics* (New York: Wiley)

[3] Hecht E 2017 *Optics* (Harlow: Pearson Education)

[4] Yariv A and Yeh P 1984 *Optical Waves in Crystals* (New York: Wiley)

[5] Fowles G R 1989 *Introduction to Modern Optics* (New York: Dover)

[6] Mahajan V N 2014 *Fundamentals of Geometrical Optics* (Bellingham, WA: SPIE Press)

[7] Mahajan V N 1998 *Optical Imaging and Aberrations, Part I: Ray Geometrical Optics* (Bellingham, WA: SPIE Press)

Chapter 3

LED die-level light extraction optics

In developing solid-state lighting based on LED, the light extraction analysis of LED dies is a crucial aspect of designing and manufacturing high-brightness LEDs. For such a complicated component system, an effective analysis method is required before the weighted relationship among these factors that would impact the light extraction efficiency (LEE) can be quantified precisely. This chapter mainly introduces the optical analysis techniques to study and enhance the LEE of LED dies. These techniques are established on the Monte Carlo ray-tracing method, and build applicable optical models for LED light extraction simulation setup and analysis. The conclusions from the LEE analysis will be used as a reference and guideline for future LED die design and manufacturing.

3.1 Challenge in LED light extraction efficiency

To determine the root causes that affect the LED light extraction, we must first understand the nature of some optical phenomena. Three main optical loss factors limit the LEE of LED dies. First, total-internal-reflection loss is caused by the high refractive index medium of the LED die itself, which is the key to the most significant impact on LEE. Second, Fresnel reflection loss is caused by the refractive index difference between the two mediums. Third, absorption loss is induced by its own semiconductor materials. In this section, we will discuss the mathematical description of these optical loss factors.

3.1.1 Total internal reflection loss

When a light ray propagates from a medium of refractive index n_1 to another medium of refractive index n_2, the incident light and refracted light will follow Snell's law (also referred to as the law of refraction):

$$n_1 \sin\theta_1 = n_2 \sin\theta_2, \tag{3.1}$$

where θ_1 is the incidence angle and θ_2 is the refraction angle. Therefore, when the light ray propagates from a dense medium into a light medium (i.e., $n_1 > n_2$), the

doi:10.1088/978-0-7503-2368-0ch3

refraction angle will increase as the incidence angle increases. When the refraction angle is larger than 90°, all the light will be reflected into the dense medium to cause a total internal reflection. The incidence angle that makes the refraction angle equal to exactly 90° is called the critical angle. Based on the Snell's law, it is derived that the critical angle θ_c is

$$\theta_c = \sin^{-1}(n_2/n_1). \tag{3.2}$$

Therefore, only light within an angle smaller than the critical angle stands a chance to be extracted. Otherwise, when the incident angle is too large, the light will be trapped inside the LED die due to total internal reflection. For the top surface of the LED die, all incident light smaller than the critical angle is equivalent to forming a light-escape cone (LEC), as shown in figure 3.1. To estimate how much light can escape from an LED die with refractive index n_1 to the external environment with refractive index n_2, an isotropic point light source is used to calculate the ratio of an LEC to the entire sphere. First, the solid angle of an LEC is calculated

$$\Omega_{LEC} = \int_{\theta=0}^{\theta_c} \int_{\phi=0}^{2\pi} \sin\theta \, d\theta d\phi = 2\pi(1 - \cos\theta_c) = 4\pi \sin^2\frac{\theta_c}{2}, \tag{3.3}$$

where θ and ϕ are the angle of latitude and longitude, respectively. In particular, for an isotropic light source, the radiation intensity I is independent of the angles θ and ϕ. And its total optical power P_{all} is $4\pi I$, because the light passes through an entire sphere with a solid angle of 4π steradians. Similarly, the optical power P_{esc} contained in the LEC can be described as $I \times \Omega_{LEC}$. Consequently, the ratio of light that can escape from one LED surface would be

$$P_{esc}/P_{all} = \sin^2\frac{\theta_c}{2}. \tag{3.4}$$

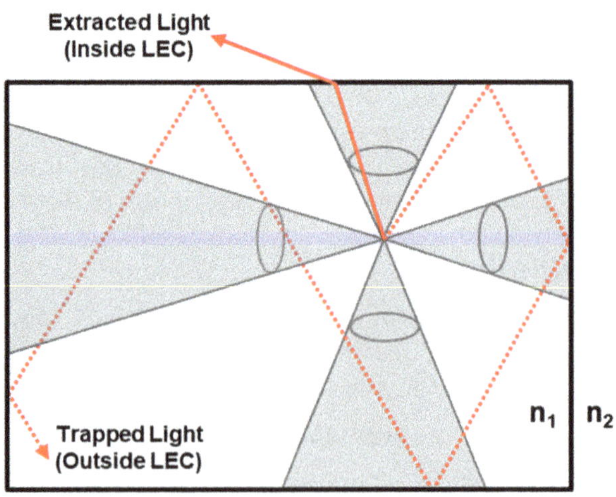

Figure 3.1. LED light extraction issue and its LEC.

Since LED semiconductor materials usually have a high refractive index, the critical angle is never huge. So, the sine term of formula (3.4) can be expanded to a power series, and the higher-order terms can be omitted. With that said, equation (3.4) can be simplified

$$P_{esc}/P_{all} \approx \left(\frac{\theta_c}{2}\right)^2 - \frac{\left(\frac{\theta_c}{2}\right)^4}{3} + \frac{2\left(\frac{\theta_c}{2}\right)^6}{45} + \dots \approx \frac{1}{4}\theta_c^2. \tag{3.5}$$

Based on equation (3.2), θ_c is also approximated to n_2/n_1 and finally arrives at

$$P_{esc}/P_{all} \approx \frac{1}{4}\left(\frac{n_2}{n_1}\right)^2. \tag{3.6}$$

This result means that there will be a certain amount of energy loss due to the limited LEE when light propagates from a dense medium to a light medium.

According to equation (3.6), when the refractive indexes of n_1 and n_2 are nearly equal, there could be less boundary to stop light from escaping. In contrast, when the difference of refractive index of the two media around the interface becomes significant, the light escape percentage will be near zero. A further explanation is that the light emitted from the active layer must be within the LEC to leave the LED. The light outside the LEC will be reflected back and forth inside the LED and will be absorbed by the material. Therefore, the intuitive idea is to increase escaping faces to increase LEE effectively [1–26]. For example, when the substrate absorbs light in a thin LED die, only one escaping face, the top surface, is allowed. In a thick die, the additional side faces increase the total escaping face area. If finally replaced with a transparent substrate, six escaping faces can be formed to increase the LEE to a certain maximum. In addition, to reduce the loss caused by the total internal reflection, the refractive index of n_2 could be increased via packaging from 1(air) to 1.5(silicone) so that a larger LEC is helpful to light escape. In the case of GaN with a refractive index of 2.42, with packaging, the critical angle increases from 24° to 39°, and the ratio of the LEC increases from 4.3% to 9.8%, a 2.3-times boost of the LEE. Moreover, although the LEC remains unchanged, die shaping and surface treatment could change the propagation direction of the light inside the LED die to increase the possibility of escape. Details will be discussed in sections 3.3–3.5.

3.1.2 Fresnel reflection loss

When light passes through an interface with two kinds of medium, a part of the light transmits and enters the new medium while the other part of the light is reflected back to the original medium. And how much is the transmittance and reflectance of the interface? The Fresnel's equations of the electromagnetic theory thoroughly answer the questions raised above. To entirely account for transmittance and reflectance, one has to consider polarization described in section 2.7, one of the characteristics of light. As the light generated by the LED is non-polarized, the reflectance can be obtained by averaging the two:

$$R = (R_s + R_p)/2. \tag{3.7}$$

In the ideal situation where the absorption is not considered, the transmittance T can be derived by the reflectance based on the conservation of energy,

$$T = 1 - R. \tag{3.8}$$

Fresnel reflection is an inevitable loss. Many photons radiated from the active layer pass through the interface of each layer. When the photons are extracted, the energy will be partially attenuated due to Fresnel reflection loss. For example, when light enters the air vertically from GaN, only 83% of the energy can pass through. If an encapsulating lens with a refractive index of 1.50 is used, the light transmittance can be as high as 94%. In addition, applying a layer of anti-reflection (AR) coating between the LED die and the external can also effectively reduce Fresnel reflection loss, as shown in figure 3.2. If the thickness (d) of the AR coating and its refractive index (n_{AR}) can meet the following conditions, the Fresnel reflection loss caused by normal incidence can be completely eliminated

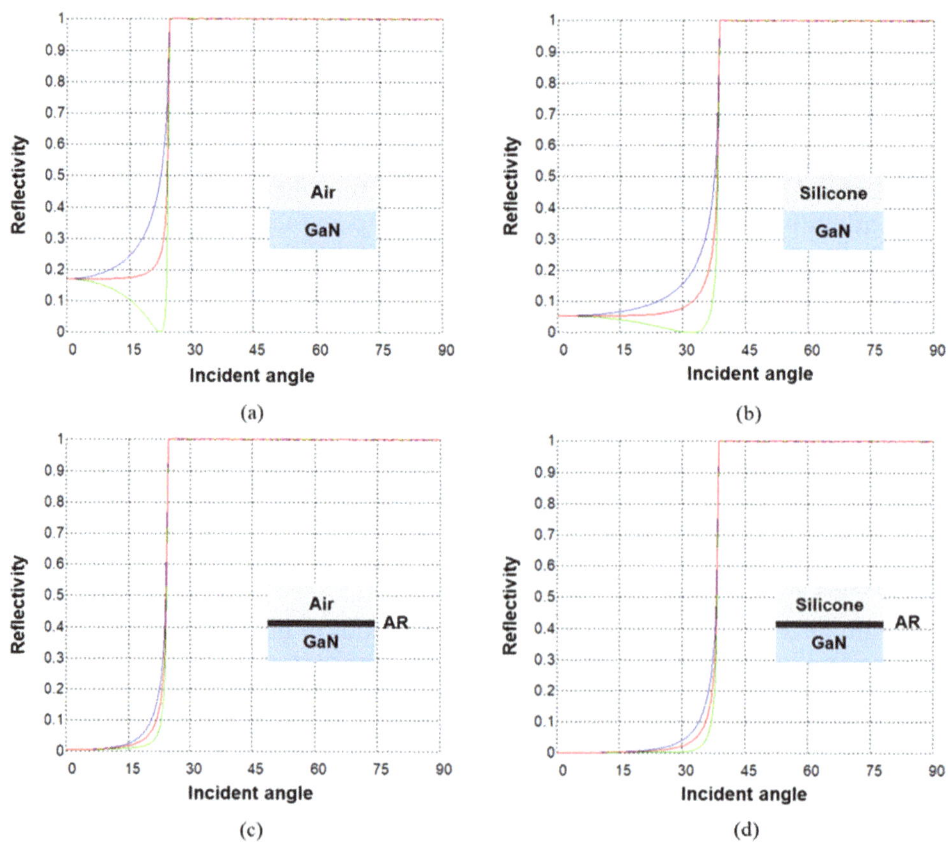

Figure 3.2. The Fresnel reflection simulation under different incident angles from 0° to 90°: (a) GaN/air interface; (b) GaN/silicone interface; (c) GaN/AR/air interface; (d) GaN/AR/silicone interface. (blue: s-polarization, green: p-polarization, red: average).

$$d = \lambda/4n_{AR}, \tag{3.9}$$

$$n_{AR} = \sqrt{n_1 n_2}. \tag{3.10}$$

3.1.3 Material absorption loss

When photons are transmitted in a medium that has the absorption characteristic, the energy will be lost due to absorption in the medium. In particular, when the energy of the incident photons is greater than or equal to the energy gap of the medium, the reduction of energy will be more noticeable. Thus, in an LED semiconductor device, the photons that cannot be extracted will eventually be absorbed by the active layer, substrate, electrodes, or mirror. If the photons are absorbed by the substrate, electrodes, or mirror, it is similar to the non-radiative recombination, and photons are transferred to heat. If the active layer absorbs the photons, some electrons will jump to the excited state to form the electron–hole pair again. A part of the electron–hole pair will radiate photons again. The ratio of the radiation is related to the internal quantum efficiency. However, the internal quantum efficiency of an LED cannot reach 100%. As a result, the intense absorption by the active layer will still cause the loss of the LEE of LED. To reduce the re-absorption of light by the active layer, the active layer is made very thin by the way of the multiple-quantum well (MQW) so that the absorption of photons by the active layer will become less [27, 28].

Regardless of what type of light absorption is, from the perspective of energy, absorption causes decay of the intensity of light. The degree of decay can be described by the famous Bouguer–Lambert–Beer law of absorption. Let us presume that the light with the intensity I_0 passes the absorption medium of the thickness L, the light intensity decays to I at a distance z and to $I–dI$ at the distance $z+dz$. In other words, the light intensity decreases by dI within the distance dz. According to basic physical consideration, the flux absorbed by dz should be proportional to Idz, i.e.,

$$-dI = \alpha I dz, \tag{3.11}$$

whereas α in the equation is defined as the 'absorption coefficient' of the medium, meaning the ratio of the intensity of the light absorbed by the medium at the unit distance. Integrating equation (3.11) along the travel path and assuming α is a constant in a uniform medium, we can obtain

$$I(z) = I_0 e^{-\alpha z}. \tag{3.12}$$

This is the description of the Bouguer–Lambert–Beer absorption law. This formula explains that as the absorbing medium gets thicker, the light intensity decays exponentially. The larger the absorption coefficient is, the more intense the absorption by the medium will be. Whether the absorption coefficient is large or small depends on the characteristics of the media, which are usually the function of the wavelength. In addition, as shown in figure 3.3, the light propagation trajectory

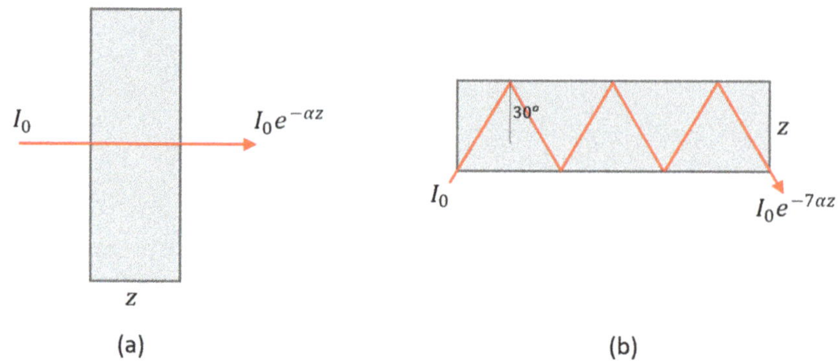

Figure 3.3. The absorption loss caused by light incident on the medium with absorption coefficient α in (a) single pass and (b) internal multiple reflections.

in the LED die is random, and there is multiple back and forth. If the photons are not absorbed by the medium, they must hit the boundaries of the LED dies multiple times. More hitting times increase absorption possibility, thus the medium absorption loss will be an important factor in LEE. Also, the LEE cannot simply explained with a notion of one-shot Fresnel loss and/or absorption loss on a certain boundary.

3.2 Effective solutions for LEE enhancement

As mentioned in the previous section, due to the physical limitations of the LED itself, a large part of the photons is confined inside the die, eventually causing low LEE. Therefore, overcoming the problem of light extraction is one of the key points to improve the overall efficiency of LEDs. Over the past two decades, many different solutions to the LEE problem have been proposed. This section will comprehensively review these general solutions currently used to boost LEE.

3.2.1 Die shaping

The general standard LED die appearance is a cube. According to its geometry shape, when the light suffers from total internal reflection at one face of the LED, it will be reflected again by the next face, and in the end, the light will remain trapped inside the LED die. To allow more light to fall into the light-escape cone, one can shape the die geometry to change the light path inside the LED die.

Krames *et al* first used the method of die shaping to resolve the problem of the cubic-shaped geometry [29]. The process of making it is to replace the light-absorbing GaAs substrate of the AlGaInP LED die with a transparent GaP substrate and cut the die into a shape similar to a 'truncated inverse pyramid (TIP)'. This die-shaping LED combines a transparent substrate and die shaping to reduce light absorption by the substrate while allowing it to escape from the die more easily. As a result, the EQE was found to be improved from 30% to 55% with the approach.

OSRAM later also successfully applied similar structures on blue LED dies using SiC substrate [30]. Originally, the LEE of the cube-shaped LED was reported 25%.

When the die was shaped, the LEE could reach 52%. If the die was flipped and a high-reflection layer was coated on the bottom, the LEE could even go as high as 60%.

In theory, the method of die shaping can also be utilized on LEDs based on a sapphire substrate, but due to the sapphire substrate being extremely hard and corrosion-resistant, it is difficult to alter its geometrical structure. For this reason, Thompson *et al* proposed to remove the sapphire substrate and replace it with an n-type ZnO substrate with a truncated inverted hexagonal pyramid shape [31]. Here ZnO provides high transparency at blue bands and allows current to conduct in it. Furthermore, the essential point is that ZnO can easily change its shape through selective etching. The literature points out that the pyramid shape can shorten the path length within the LED die while minimizing the possibility of total internal reflection or material absorption.

3.2.2 Surface roughening/texturing

Texturing or roughening the top surface of LED dies is now a common technique in improving the LEE. As early as 1973, Bergh and Saul from Bell Telephone Laboratories proposed the concept of rough surface to destroy the total internal reflection in electroluminescent diodes with a high refractive index [32]. It was not until 1993 that Schnitzer *et al* first used the natural lithography process to roughen the surface of GaAs LEDs to increase its EQE from 9% to 30% [33]. It is also applicable for GaN-based LEDs. However, directly roughening through the etching thin top P-GaN layer might cause electrical deterioration. To change the epitaxial growing condition to control surface roughness of P-GaN is a more feasible way for now [34–37]. Another way is using the substrate replacement technique to replace sapphire with a material having higher thermal conductivity. The thicker N-GaN layer happens to be flipped to the top surface, which is the more readily etched surface. Due to the new substrate of this LED die being opaque, only a thin epitaxial layer can extract light, so it is commonly called ThinGaN LED [38–41]. According to the literature, using photoelectrochemical etching technologies to form a random-distributed pattern of hexagonal pyramids on the N-GaN of a ThinGaN LED, the light intensity is 2.3 times that of a ThinGaN LED with a smooth top surface [42].

Besides roughening the top surface, roughening the side surfaces or the interface of the LED die can also further boost the LEE. Lin *et al* increased the LEE to 82% by roughening the side surface [43]. Hus *et al* further roughened the P-GaN surface, the N-GaN etched surface, and the side surfaces. They found that the LEE increased by 1.41 times compared to only performing surface roughness of the P-GaN and 2.57 times higher than LEDs without roughening. This shows that the LEE improvement becomes much more noteworthy as more surfaces are roughened [44].

3.2.3 Patterned sapphire substrate

Many works in literature point out that the GaN epitaxial layer grows horizontally on a patterned sapphire substrate (PSS) for reducing the 'dislocation', improving the epitaxial quality, and thus enhancing the internal quantum efficiency (IQE) [19, 45–50]. However, due to the apparent difference in refractive index between the sapphire substrate and the

GaN epitaxial layer, some studies report that microstructures on the sapphire substrate, just as with the surface roughening/texturing, can enhance the LEE. The two effects can both improve the overall efficiency of the LEDs. NICHIA was the pioneer to use PSS technology and ITO transparent mesh electrodes to increase the luminous efficacy of white LEDs to reach 150 lm W^{-1} [51]. In addition, many reports describe that the LEE could be further enhanced by using the periodic microstructure in the PSS. The factors include the shape, size, depth, and pitch of the microstructure. The deeper the microstructure etching is and the higher the density is, the larger the LEE [52–55].

3.2.4 Photonics crystal

Photonic crystals are a dielectric structure with a periodic distribution. When electromagnetic waves penetrate into photonic crystals, due to destructive interference at specific frequencies, the electromagnetic waves cannot effectively pass through the structures, and this phenomenon is called photonic band gap. Based on this characteristic, various optoelectronic components are proposed. Enhancement of the LED efficiency is one of the applications of the photonic crystal. Some literature points out that a photonic crystal can improve LED efficiency in contrast to other physical mechanisms [56, 57]. When the photonic crystal is used in the active layer, the spontaneous radiation rate can be increased via the Purcell effect. Moreover, the waveguide mode can be decoupled into the radiation mode. Practically, however, the photonic crystal structure could reduce the effective area of the active layers and increase non-radiative combination. When photonic crystal structures are placed on the top surface of the LED die, the diffraction by the structure could increase the LEE.

To sum up, the general ways to improve the LEE of a LED die include changing the die shape and creating regular or irregular micro/nano structures on the die surface or interface. These light extraction structures can effectively change the light path within the die to increase light extraction. Other than these, designs involving transparent electrodes, high-reflective mirrors, substrate replacement, and flip-chip packaging are also helpful for boosting LEE. We will discuss these aspects in the following sections.

3.3 LED light extraction analytical method and simulation

In 1974, Joyce and others presented the concept of the random particle for the first time to describe the spontaneous emission process [58]. In 1995, Ting and McGill applied the Monte Carlo ray-tracing method to analyze the LED light extraction [59]. The Monte Carlo method can effectively simulate the random generation of photons in the active layer shown in figure 3.4(a). Furthermore, combined with ray-tracing calculation, one can trace the light path inside the LED and estimate how much light can escape from the die, as shown in figure 3.4(b).

The Monte Carlo ray-tracing method, however, is unable to explain the wave nature of light, such as the diffraction, waveguide mode, evanescent wave, and other effects. Thus, some studies would turn to the wave theory to simulate light propagation inside LEDs [60–64]. Compared to the ray-tracing method, the wave-based approach has to process complex wave equations and greatly increase

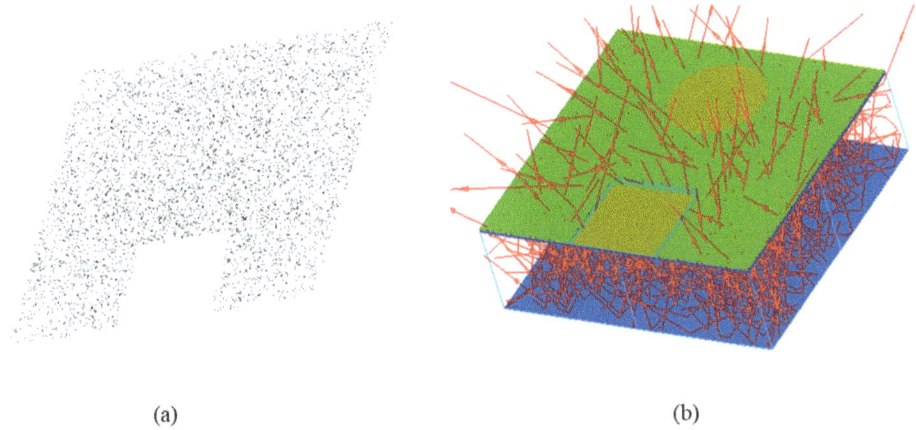

Figure 3.4. LEE simulation with the Monte Carlo ray-tracing method. (a) the random distribution of the photons; (b) the light path of the photons inside the LED.

computing time. To make the wave simulation execute smoothly, the simulation needs to be confined in a local volume, the structure needs to be simplified or photon recycling effect is ignored. As a result, the LEE simulation is not accurate. Therefore, in most cases, the Monte Carlo ray-tracing method is still the most suitable simulation core for the LEE calculation.

There are four main steps to use the Monte Carlo ray tracing method to simulate the LEE of LEDs, as described in the following: The first step is to establish the actual LED model, including setting up the parameters such as the geometry of the components, the properties of the materials and the surfaces. To match the actual physical behaviors, we regard the photons generated in the active layer as the collection of many arbitrary point light sources. Thus the second step is to turn on the random-distributed point sources across the active layer. The third step is to trace the light rays upon transmission, reflection, absorption and scattering. The tracing of a ray does not stop until the photon escapes from the die or is absorbed. The former contributes to the rate of LEE and the latter does not. The final step is to analyze the LEE.

When the Monte Carlo method is applied to explain a problem, sampling rate is a key factor to decide how precise the simulation is. A higher photon number will make the simulation more reliable or accurate, but this requires a considerably long period of computing. Therefore, before conducting the analysis, we have to consider balancing the precision and computation time. In the Monte Carlo ray-tracing simulation, the parameters that could affect the computation time and precision mainly include the number of ray generations, the number of ray intersections, and the number of ray splits. To understand the extent of the impact, two LED structures are used for the simulation test, i.e., GaN-based LEDs with and without patterned sapphire. Generally, the number of ray generations required for an optical simulation is at least 1 million rays. Depending on the extent of the split of the photons, the number of rays in the end may be around 5 million to 80 million at least. Ray tracing performed on such a bulk amount can help obtain precise results of the analysis. In terms of the study on the LEE, figure 3.5 shows the relationship between the LEE and the number of ray generations. The maximum intersection time is set at 1000, while the split time is set at 1. The result shows that the LEE needs only around 1000 rays to converge. The same theory applies to what is illustrated in figures 3.6 and 3.7. A stable value is arrived at when the intersection

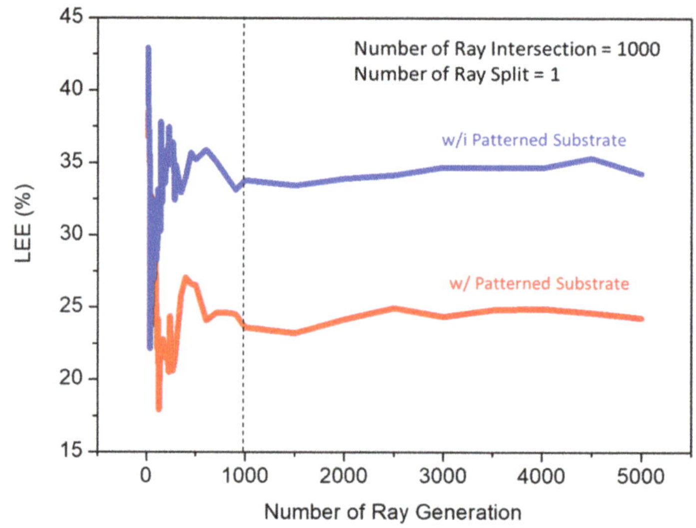

Figure 3.5. The relationship between the LEE and the number of ray generations.

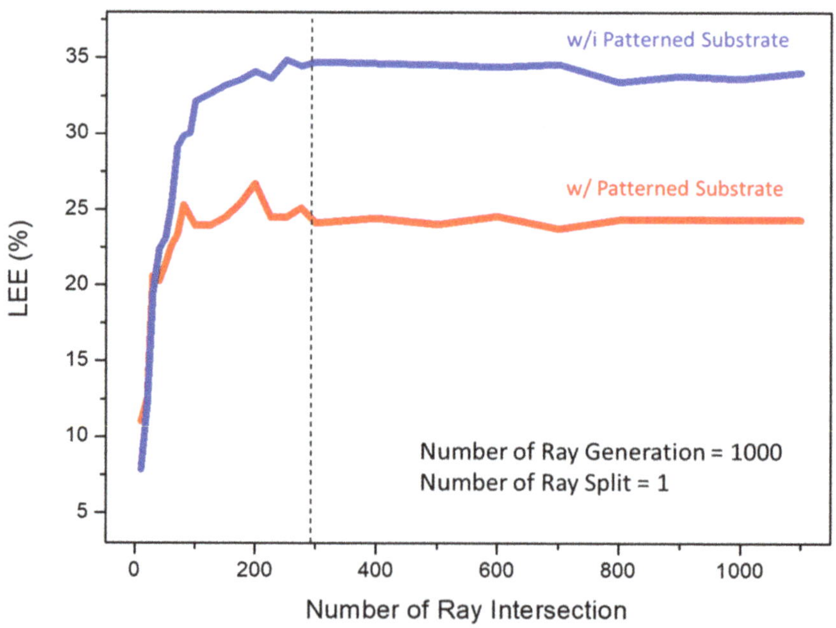

Figure 3.6. The relationship between the LEE and number of ray intersections.

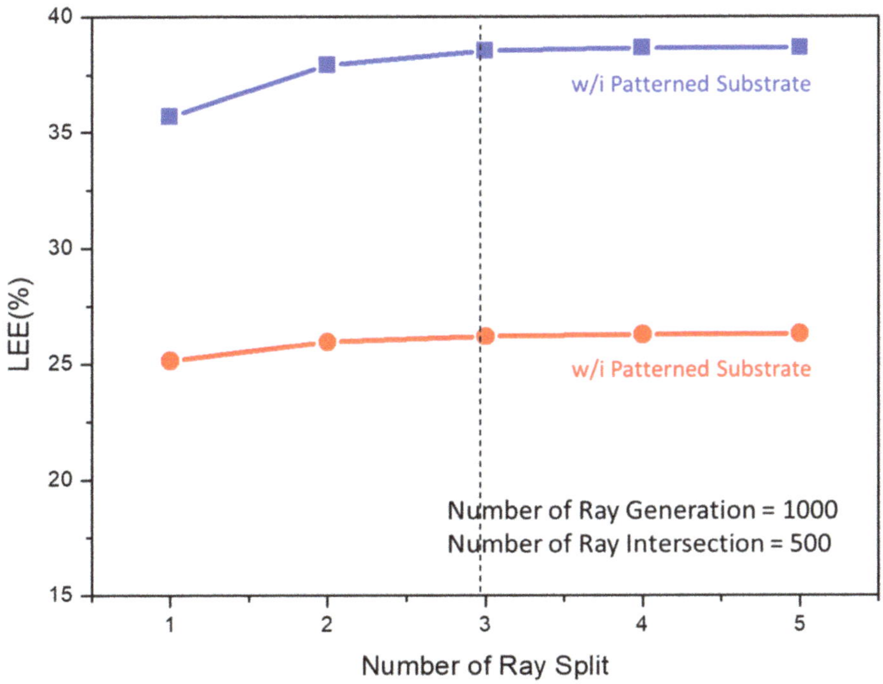

Figure 3.7. The relationship between the LEE and number of ray splits.

time and the split time reach 300 and 3, respectively. This shows that stable and precise results of the analysis on the LEE can be obtained without requiring so much ray generation and split. This is primarily because the calculation of the LEE is merely the integral of the energy of photons. It does not consider issues such as spatial resolution. Therefore, the Monte Carlo method is pretty suitable for optimizing the structure of the LEE, but please note that the simulation must be in the scope of geometrical optics.

3.4 Case studies of LEE simulation

Next in this chapter, seven different GaN-based blue LED dies will be used for LEE simulation, including three basic and four advanced structures, as shown in figure 3.8. The difference between the three basic LED dies is that they have different substrates: sapphire, SiC, and silicon, respectively. Sapphire and SiC are growth substrates of GaN-based LED dies which are transparent in the visible light region. In order to improve heat transfer performance, silicon or metal substrates with higher thermal conductivity will be used instead of sapphire. Since silicon and metal are generally opaque to visible light, it is often necessary to coat the substrate with a highly reflective layer in advance. Then, the four advanced LED dies have additional light-extraction structures, as mentioned in section 3.2, including die shaping of GaN-on-SiC LED, surface texture of GaN-on-sapphire LED, patterned substrate on GaN-on-sapphire LED, and surface texture of GaN-on-silicon LED.

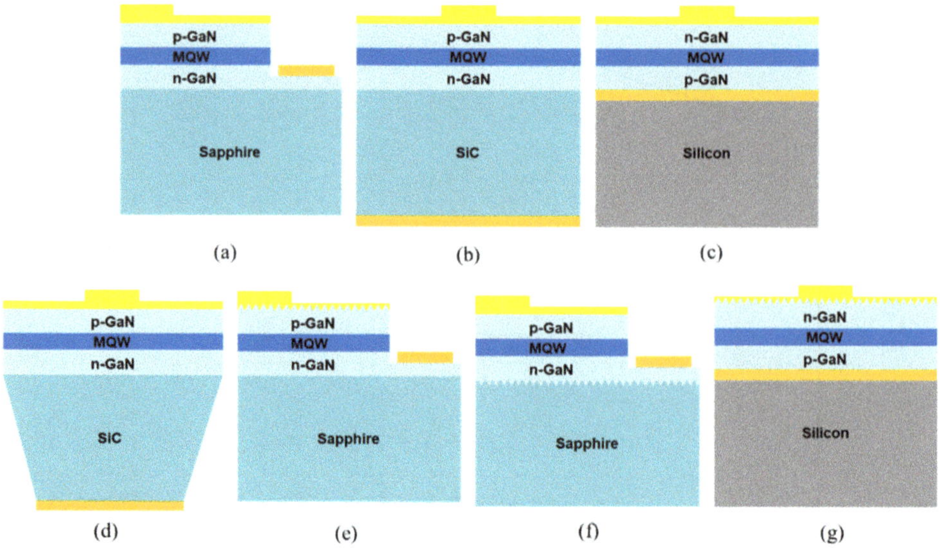

Figure 3.8. The simulated structures of the GaN-based LED die. (a) GaN-on-sapphire, (b) GaN-on-SiC, (c) GaN-on-silicon, (d) die shaping (SiC), (e) surface texture (sapphire), (f) patterned substrate (sapphire), and (g) surface texture (silicon).

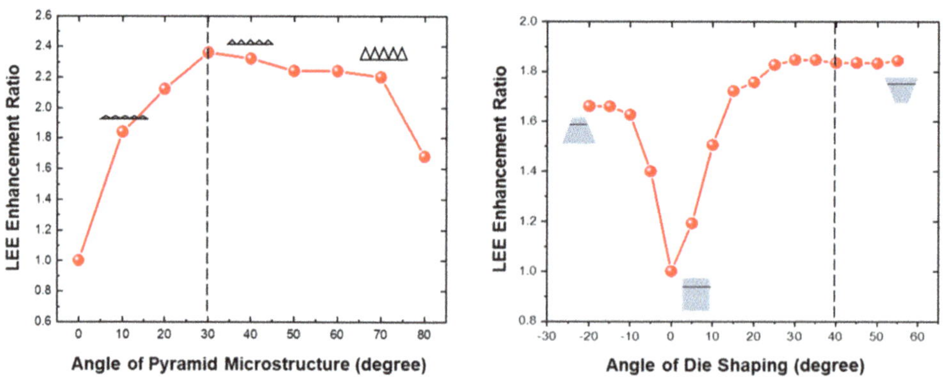

Figure 3.9. The optimization of the light extraction structures.

Figure 3.9 shows the optimization of these light extraction structures, from which a 40-degree cut for die shaping and 30-degree pyramid microstructure for surface texture and pattern substrate are applied for subsequent case studies. The epitaxial layer of all LED dies in the simulation includes the GaN buffer layer, n-GaN layer, InGaN/GaN MQW active layers, AlGaN barrier layer, p-GaN layer, and the ITO transparent electrode. The parameters of each layer, such as thickness and the refractive index, are listed in table 3.1 [65].

The first case study is to analyze the effect on LEE by absorption of the active layer. In theory, the active layer not only generates photons but also absorbs the photons passing through it. For this reason, strong absorption by the active layer is

Table 3.1. GaN-based LED die simulation parameters.

	Thickness (μm)	Refractive index
ITO	0.2	1.90
p-GaN	0.1	2.45
AlGaN	0.1	2.42
MQW	0.1	2.54
n-GaN	4	2.42
Sapphire	100	1.78
SiC	100	2.70

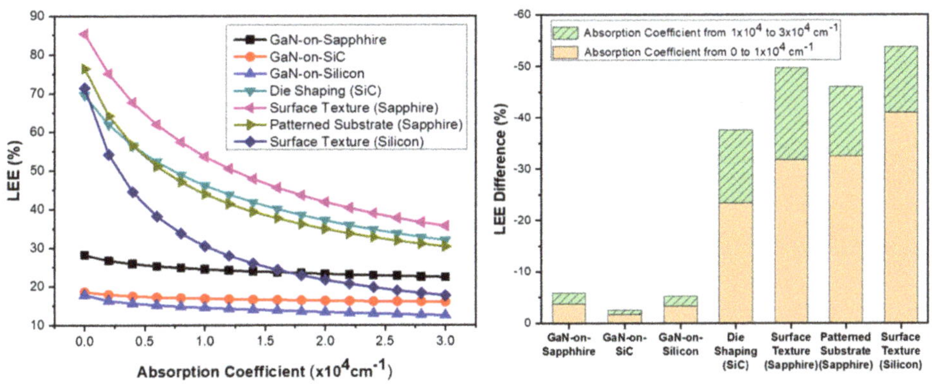

Figure 3.10. The impact of the absorption coefficient of the active layer on the LEE.

an essential factor not to be ignored when analyzing the LEE. Figure 3.10 shows the simulation of the LEE as a function of the absorption coefficient of the active layer. Obviously, the increase of the absorption coefficient of the active layer results in the drop of the LEE, especially for LEDs with light extraction structure. The reason is that an LED with a light extraction structure usually requires a relatively large number of times of photon recycling to effectively increase the LEE. The photon recycling here refers to the number of multiple reflections of light inside the LED cavity and we refer to this as 'cavity photon recycling'. Basically, the more photon recycling there is, the longer the distance will be for the photons to travel inside the LED. So the chance of the photons being absorbed by the active layer will increase. Figure 3.11 shows the relationship between the accumulated number of times of cavity photon recycling and the LEE under different absorption coefficients of the active layer. The analysis results show that the absorption effect poses a greater threat to LEDs with light extraction structures. Therefore, in the broader sense, reducing material absorption is also an indispensable key factor in enhancing LEE.

Next, the impact of the LED die size on LEE is analyzed, as shown in figure 3.12. Traditionally, for low-power LEDs, the die area is about 0.3 × 0.3 mm^2, and the standard operating current is around 20 mA. To enable LEDs to operate at high

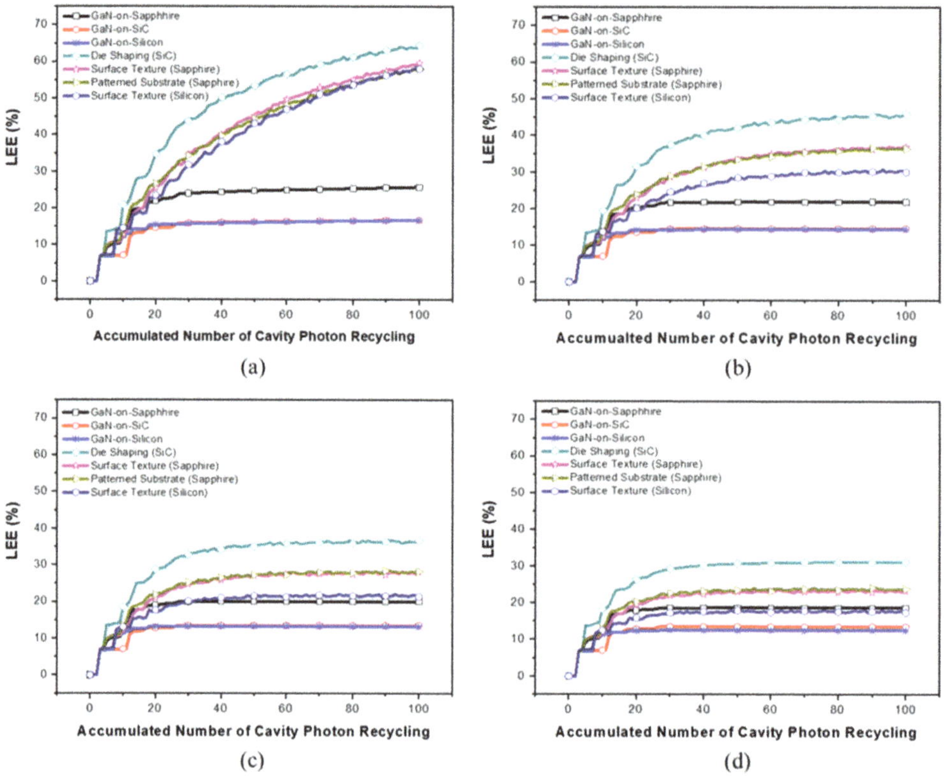

Figure 3.11. The relation between the accumulated number of times of cavity photon recycling and LEE when the absorption coefficient of the active layer is (a) zero, (b) 1×10^4 cm^{-1}, (c) 2×10^4 cm^{-1}, and (d) 3×10^4 cm^{-1}.

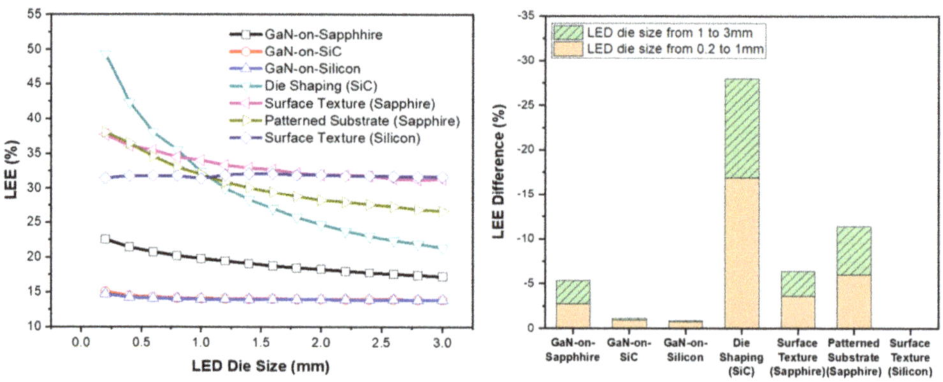

Figure 3.12. The impact of the LED die size on LEE.

power, the die size must be large enough to withstand higher input current. Therefore, for LEDs with an input power of 1 Watt or more, the die area needs to be increased to 1×1 mm^2. The analysis results show that only GaN-on-silicon LED with surface texture can maintain LEE when the die size becomes large.

In contrast, LEE decay is most significant in GaN-on-SiC LED with die shaping. Such LEE changes are mainly due to the fact that photons in large-sized LED dies need to travel a longer distance before reaching the side. This will increase the light absorption inside the LED, thereby reducing the LEE on the side of the LED. Therefore, for die shaping and patterned substrate, since most of the light is extracted from the side of the LED die, increasing the die size has a relatively significant impact on LEE. As a result, for large-size LEDs, increasing the ratio of LEE on the top surface of the die is the most direct and effective solution.

Another dimension that needs to be addressed in LED die geometry is thickness. The epitaxial layer is usually very thin, a few microns, so the thickness of the LED die is mainly from the substrate. Figure 3.13 shows the effect of transparent substrate thickness on the LEE. The results show that the LEE will be significantly improved with the increase of substrate thickness in the GaN-on-SiC LED with die shaping. However, the LEE of GaN-on-SiC without die shaping does not improved as the substrate becomes thicker, because the SiC has a high refractive index, resulting in a small LEC. In addition, for GaN-on-sapphire LED with patterned substrate, when the substrate thickness increases from 10 to 100 μm, the LEE improves, but if the thickness continues to increase, the LEE is close to saturation. The same results were observed for GaN-on-sapphire LEDs with and without surface texture. As a result, when it comes to the design of the transparent substrate, it is not true that the thinner the substrate is, the better. Instead, one has to consider the changes of the LEE and the impact caused by the internal thermal resistance within the device at the same time.

The Monte Carlo ray-tracing method is undoubtedly an essential tool for analyzing, designing, and optimizing the LEDs. It has been verified that it can effectively describe the light-emitting characteristics of the active layer and the transmission behavior of photons in the LED die, and then perform quantitative analysis of LEE. However, pure ray-tracing simulation only considers the refraction, reflection, and absorption, but does not take into account the complex effects such as active layer re-radiation, current crowding, and microstructure diffraction. Therefore, to further explore the impact of these factors on LEE, it is necessary

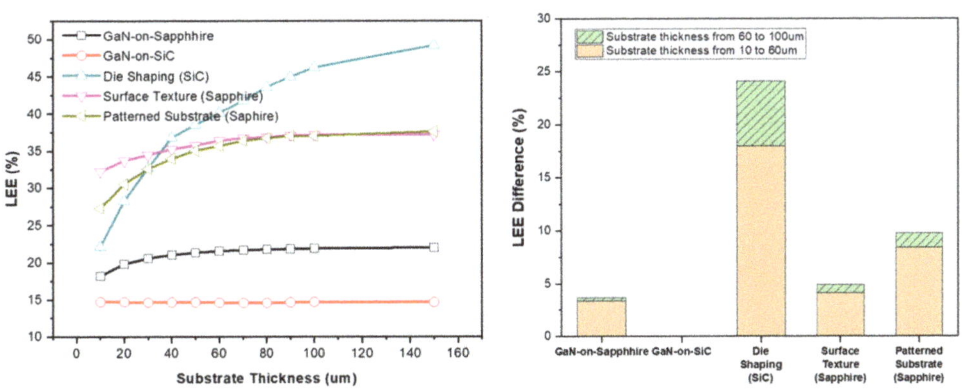

Figure 3.13. The impact of the transparent substrate thickness on LEE.

to build additional physical models to simulate these characteristics and integrate them into the Monte Carlo ray-tracing method. These models will be introduced in the following.

3.5 Quantum photon recycling mechanism on LEE

When photons are trapped in the LED die and unable to escape, part of the photons will re-enter the active layer. The active layer absorbs these photons, excites a portion of electrons and holes, and has the chance to recombine again to generate new photons. In this way, these photons are repeatedly absorbed and emitted. To not conflict with the term 'cavity photon recycling' mentioned in section 3.3, we refer to this process as 'quantum photon recycling' based on the electron–photon quantum interaction. This effect could be significant for LEE. This section will discuss the way to perform quantum photon recycling based on the Monte Carlo ray-tracing method.

As shown in figure 3.14, the detailed simulation process mainly includes several steps: (1) The photons begin to be ray-traced inside the LED. Some photons escape from the die, while some other photons are absorbed by the active layer or by other materials; (2) the location at which the photons are absorbed by the active layer and the related energy are recorded. As we know, new emission photons are emitted again from the location of the active layer. Thus, the quantum photon recycling occurs. Here, we define a factor of quantum photon recycling rate (Q_{PRR}) to describe the re-emission rate of an absorbed photon, which could be directly proportional to the IQE. Moreover, the distribution of the new radiation photons is related to the location of the absorbed photons, and their propagation direction is different from the original one before absorption; (3) return to the first step and start tracing and recording the new photons again until the LEE value becomes stable; and (4) finally, sum up the LEE of each photon recycling to have the final LEE. The formula is as follows:

$$\begin{aligned}\eta_{LEE} &= \eta_{1\mathrm{th}} + (Q_{PRR} \times \alpha_{1\mathrm{th}}(x, y, z)) \times \eta_{2\mathrm{nd}} + (Q_{PRR} \times \alpha_{2\mathrm{nd}}(x, y, z)) \times \eta_{3\mathrm{rd}} + \ldots \\ &= \eta_{1\mathrm{th}} + Q_{PRR} \times \sum_{i=1\mathrm{st}}^{i=n\mathrm{th}} \alpha_{i\mathrm{th}}(x, y) \times \eta_{(i+1)\mathrm{th}}, \end{aligned} \quad (3.13)$$

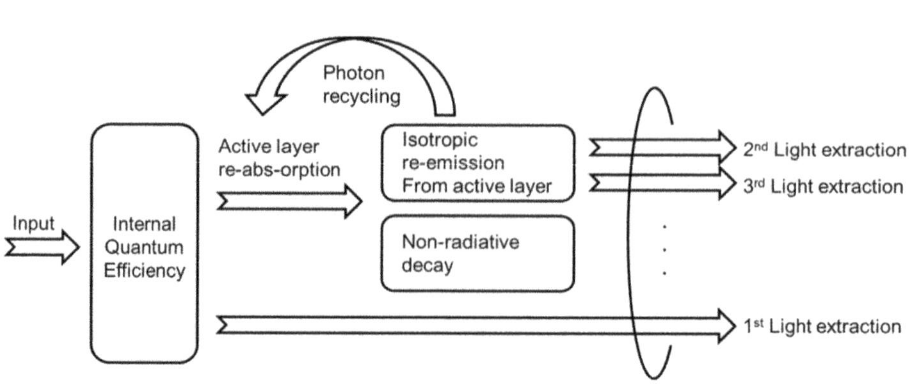

Figure 3.14. Mechanism diagram of the quantum photon recycling in LED.

whereas $\eta_{(i+1)\text{th}}$ is the LEE in the ith photon recycling, and $\alpha_{i\text{th}}(x, y, z)$ is the photon energy distribution absorbed by the active layer during the ith photon recycling. From the equation (3.13), it is easy to understand that quantum photon recycling affects LEE more dominantly in an LED die with heavy active layer absorption and large Q_{PRR}.

As explained above, the distribution of the re-emitted photons is related to the location of the absorbed photons. However, in Monte Carlo simulation, recording the 3D coordinates of all the photons will create massive data and prolong the calculation time. Therefore, one can split the active layer into multiple layers of pixels to record the location of the absorbed photons. Every time a photon passes one pixel on a layer, calculations will be performed on the pixel to record the energy absorbed. Figure 3.15 illustrates the recording method. Each pixel is regarded as the unit light source to re-emit photons, and all the photons are emitted in the way of spontaneous emission. In simulation, more layers and pixels will improve the accuracy of the calculation.

For GaN-on-sapphire LEDs with patterned substrate, figure 3.16 shows the cumulative LEE simulation of two absorption coefficients, e.g., 10^4 cm^{-1} and 200 cm^{-1}, under different photon recycling times. Each simulation contains nine recycles.

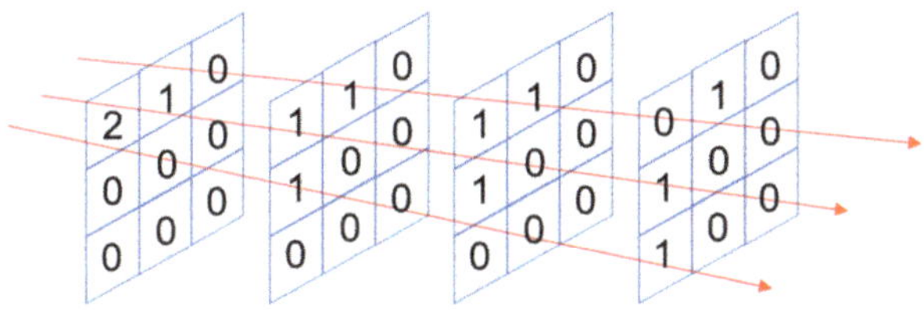

Figure 3.15. The method of recording the absorption and distribution of photons by the active layer.

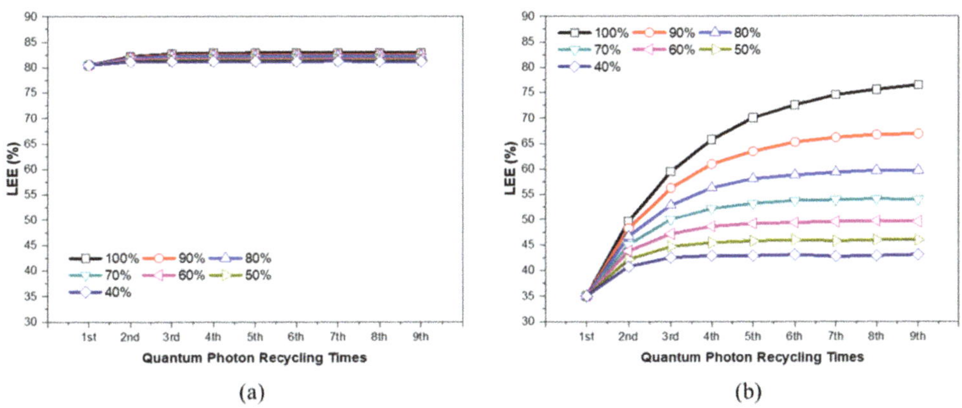

Figure 3.16. For GaN-on-sapphire LED with patterned substrate, the cumulative LEE of different quantum photon recycling rate and times when the absorption coefficient of the active layer is (a) 200 and (b) 10^4 cm^{-1}.

As shown in the figure 3.16, under strong absorption, the higher the Q_{PRR} is, the higher the photon energy re-radiated from the active layer will be. This significantly impacts the final LEE. Through the mechanism of quantum photon recycling, LEE has a chance to increase accordingly. Even if there is no light extraction structure, the enhancement of LEE may be more than double as a result. But the condition is different in the case of weak absorption. The cumulative LEE reaches saturation within three or fewer quantum photon recycles. The reason is that the active layer has poor ability to capture photons. Under such a condition, the cavity photon recycling is still an effective way to increase LEE rather than quantum photon recycling. Based on the rule of thumb, the active layer of GaN-based LEDs is closer to a weak absorption, while GaAs-based LEDs have strong absorption, so the quantum photon recycling characteristics may be quite significant. Therefore, if the LEE simulation does not consider the effect of quantum photon recycling, the actual LEE of GaAs-based LEDs will be a mystery.

3.6 Current crowding effect on LEE

LEDs need to be driven by electrodes. In addition to the electrical issues such as ohmic contact, the geometric configuration of the LED electrode should also be considered, as it will affect the injection current spreading at the active layer. For GaN-based LEDs, if insulating sapphire is used as the substrate, both positive and negative electrodes must be made on one side of the epitaxial layer. This way, it will introduce the lateral current, making it much easier to cause the current crowding effect. For high-power LEDs in particular, the current crowding effect will be more noticeable when their size is larger while being operated at high currents. If the currents are not distributed uniformly, the photon generation in the active layer will not be uniform. At the same time, it causes heat to accumulate only in certain regions, thereby attenuating the IQE and causing damage to the components. Furthermore, since most electrodes are made of opaque materials, the light will be blocked, which, in turn, reduces LEE. Therefore, LED electrode design must consider the impact on LEE, while improving the uniformity of current spreading.

To obtain a reasonable and accurate LEE, the photon distribution generated in the active layer must match the current spreading distribution. Therefore, two-stage modeling is required to integrate electrical and optical domains. The first is to determine the electronic transport properties in the LED through electrical simulations. Then, convert the electron energy distribution in the active layer into the corresponding photon distribution for ray tracing and calculate its LEE. In the first-stage electrical simulation, a feasible method is to use Poisson's equation to build a three-dimensional finite element model (FEM) to quantify the current spreading distribution in the active layer [66, 67]. The result provides an essential piece of information for the second-stage optical simulation to set up the generated photon distribution in the active layer. To make it simple to run the optical simulations, the initial energy of each photon can be set as the same, and the photon number density distribution can be proportional to the current expanded distribution, as shown in figure 3.17.

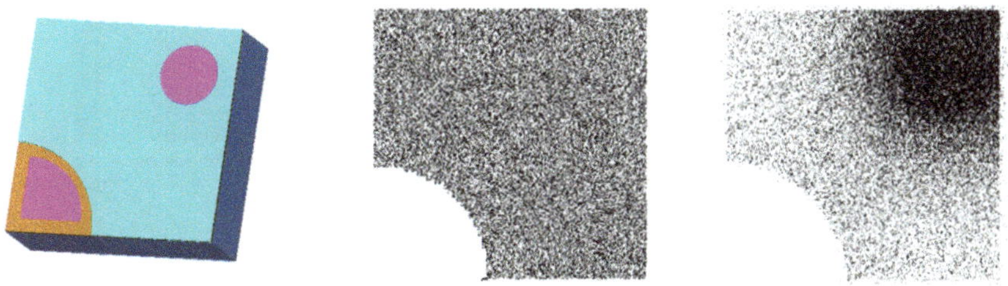

(a) (b)

Figure 3.17. The simulation of the photon generation distribution in the active layer (a) without and (b) with considering electrical current crowding effect.

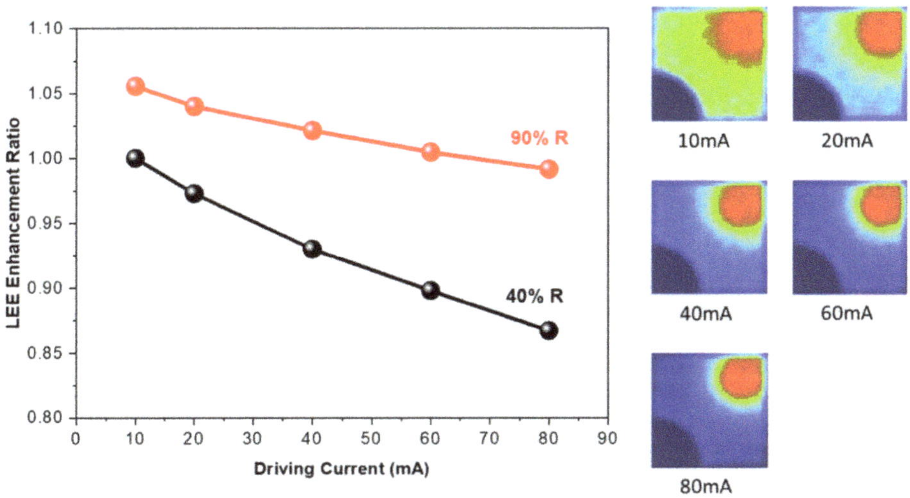

Figure 3.18. The impact of driving current and electrode reflectivity on LEE.

Take the LED in figure 3.17 as an example to simulate LEE under different drive currents, and the results are shown in figure 3.18. As the driving current increases, the current crowding effect becomes more significant, and the photon generation distribution in the active layer is relatively concentrated. The simulation results show that the current crowding and electrode reflectivity do affect LEE. Many studies have proposed various ways to overcome the problem of current crowding, such as using interdigitated or ring electrode configurations, a transparent conducting layer or current blocking layer [68–71]. Figure 3.19 shows an LED die with a rectangular structure, simulating the effect of extending the p-electrode length on LEE. It can be seen that extending the p-electrode length can indeed effectively make the photon generation distribution more uniform, and therefore help to improve the LEE, especially when the current density is high. However, most of the generated photons are still generated below the p-electrode and cannot be extracted due to electrode blocking. Therefore, if the electrode has a higher reflectivity, it should be able to

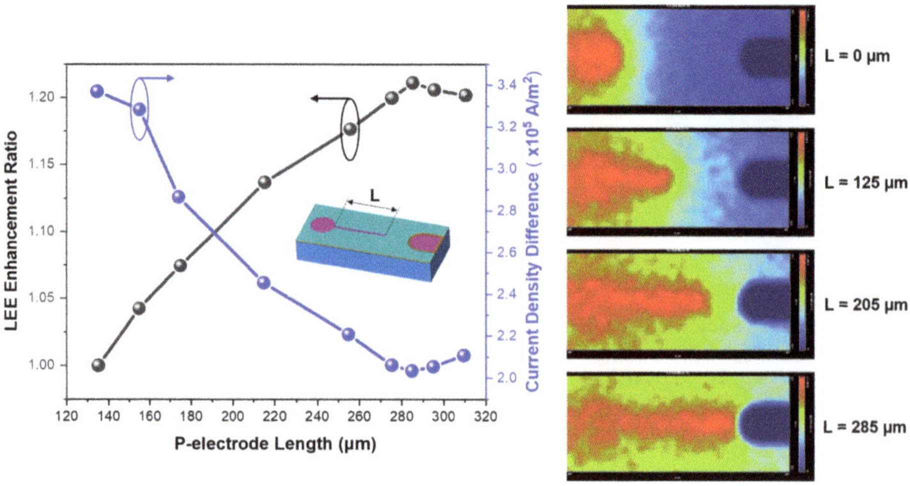

Figure 3.19. The relation of the p-electrode length to the LEE and the current density difference.

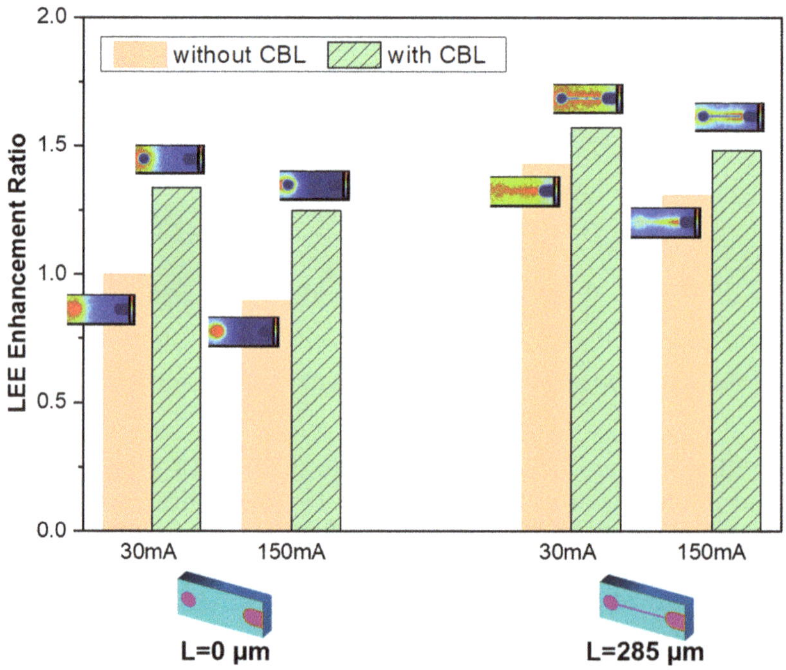

Figure 3.20. The impact of the driving currents on the LEE with and without extending the p-electrode and CBL.

minimize the loss of LEE. Another solution is to use a current blocking layer (CBL). By making an insulating layer on the bottom of the p-electrode, the current will spread to the edge of the p-electrode. The LEE analysis results show that using the CBL can further improve the LEE, as shown in figure 3.20.

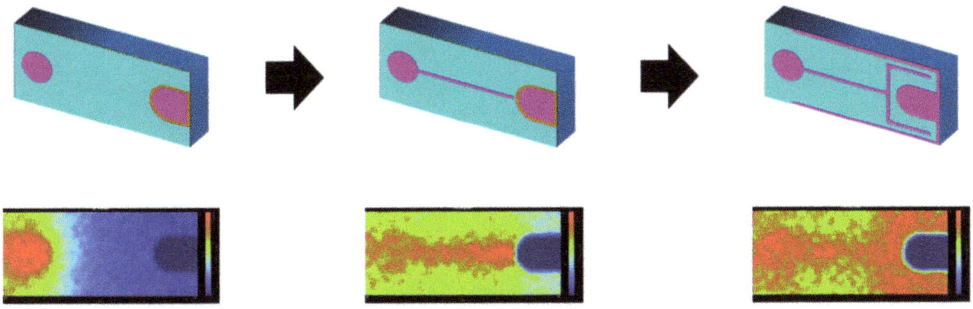

Figure 3.21. The optoelectronic integration model for LED electrode design to improve current uniformity and LEE.

In summary, as shown in figure 3.21, with this optoelectronic integration model, the impact of electrical structures on the LEE can be further clarified and quantitatively evaluated. Furthermore, this model helps with the electrode design for the purpose of enhancing the current uniformity and LEE of the LED at the same time.

3.7 Microstructure and light-scattering on LEE

To effectively improve the LEE, forming a patterned/roughened microstructure on the surface or interface of the LED die has become an essential element. Figure 3.22 shows the microstructures produced by different processes, each of which has its own uniqueness in shape, size, depth, and spacing. Therefore, questions such as which surface/interface has the most significant impact on LEE, which arrangement is more suitable for LEE improvement—periodic or random, which shape and its parameters have the best LEE or which scale—micro or nano—is better for LEE are just some examples of what we should be concerned with when designing microstructures to optimize LEE.

In theory, when light enters the randomly rough surface, it will be scattered. Different levels of roughness will bring other impacts on LEE. To accurately describe the trajectory of how light is scattered when passing through these rough surfaces, the optical properties of the microstructure, in addition to geometric parameters, must be precisely defined. In the Monte Carlo ray-tracing simulation, an optical model that simplifies the random microstructure is to treat a rough surface as a combination of many irregular surfaces. The simulated light trajectory is shown in figure 3.23. When light is reflected or refracted from these small virtual surfaces, its direction will change. Therefore, these scattered light rays can be regarded as the result of the random redirection of light.

The light-scattering model can be described using mathematical distribution functions, such as the well-known Gauss, Lambertian, Harvey, and various other models based on bidirectional scattering distribution function (BSDF) [72]. The following distribution functions can be effectively but in a simple way to describe varying surface roughness:

$$P(\theta) = K \times \cos^{1/n}(\theta), \tag{3.14}$$

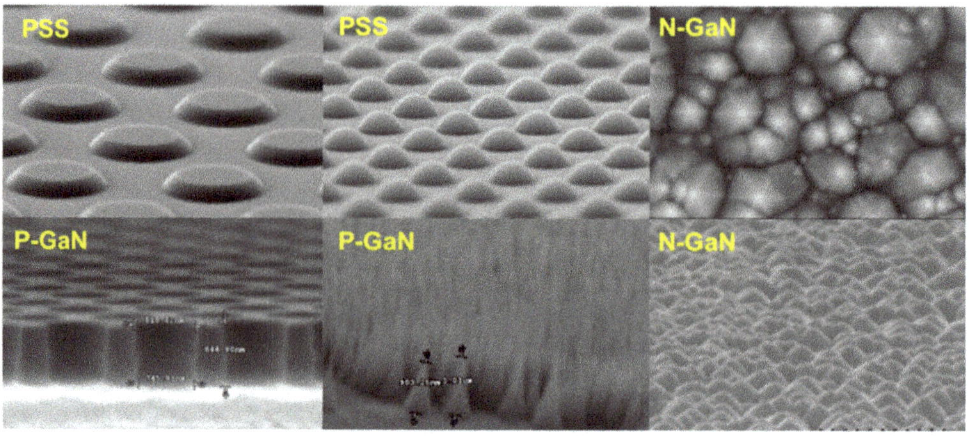

Figure 3.22. The patterned/roughened microstructure on LEDs.

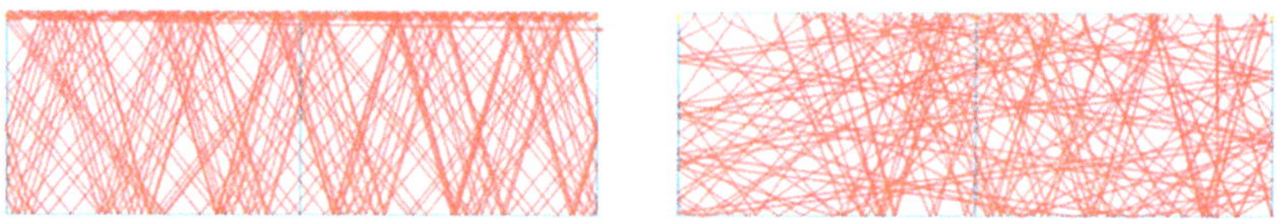

(a) (b)

Figure 3.23. Light trajectory simulation (a) without, and (b) with considering the light scattering model on GaN/sapphire interface.

whereas θ is the angle of light scattering, K is the constant, and n is used to represent the surface roughness level. The value range is 0 to 1. When $n = 1$, the surface is an ideal Lambertian scattering surface. When n decreases, the surface scattering behavior becomes weaker. When n is close to zero, the surface can be considered a plane surface, meaning not any scattering effect. Figure 3.24 takes the LED in figure 3.17 as an example to simulate the effect of surface roughness of different interfaces on LEE. The results show that the stronger the light scattering inside the LED is, the greater will be the light that can be extracted. Through light scattering, the chance of light entering LEC is increased, thus effectively enhancing LEE.

The light scattering model uses a single variable to describe the surface roughness. However, this method is only suitable for the simulation on random rough reflective surfaces. Transmission-type scattering is more complicated than reflection-type scattering because the former always has forward and backward scattering at the same time. Therefore, for periodic or quasi-periodic patterned structures used on the LED surface or interface, it is recommended to directly build the geometric shape of the actual microstructure to describe its complex light scattering behavior.

Here, the two periodic microstructure arrays, namely micropyramid and micro-lens, use ray tracing for light-scattering analysis, and their geometric shapes are

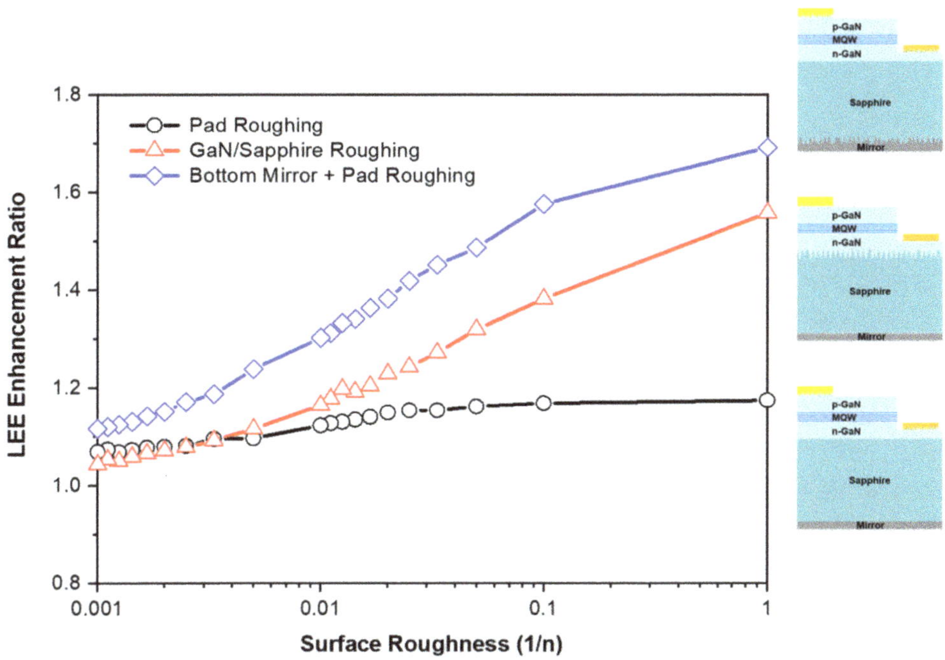

Figure 3.24. The effect of different surface roughness on the LEE.

(a) (b)

Figure 3.25. Illustration of the microstructure. (a) micropyramid array, and (b) microlens array.

shown in figure 3.25. The basic structure of the micropyramid has a square of the size of 5×5 μm^2. The arrangement is a compact square array. The height of the pyramid varies with the angle of the tiles. Another microstructure, microlens, adopts a hexagonal base and arrangement. The bottom diagonal distance is 5 μm, and the lens shape is parabolic. Therefore, the lens curvature of the lens can be expressed by the aspect ratio. Figures 3.26 and 3.27 show the two microstructures respectively, which simulate the relative energy distribution of the transmission and reflection when light is incident from the GaN material into the air. The darker the color, the stronger the energy, and vice versa. The results show that the microstructure does change the angle and energy distribution of transmitted and reflected light.

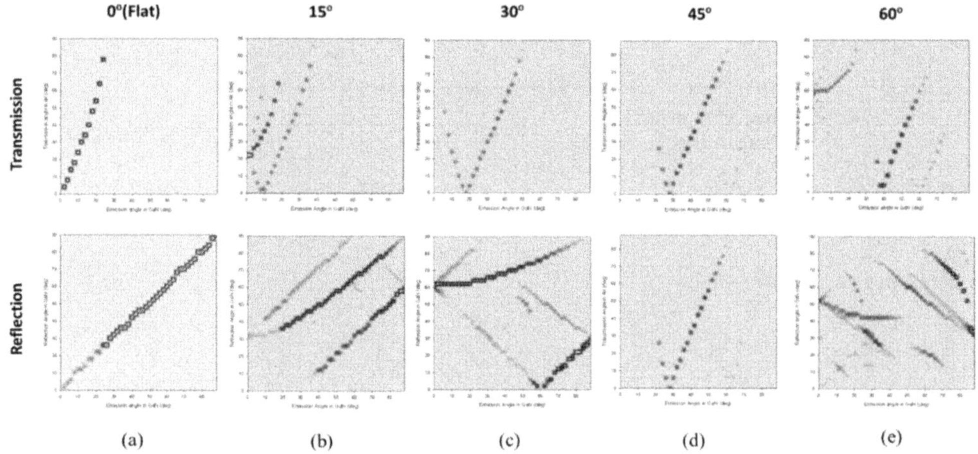

Figure 3.26. The angular and relative energy distribution of the transmission and the reflection of (a) 0°, (b) 15°, (c) 30°, (d) 45° and (e) 60° micropyramid.

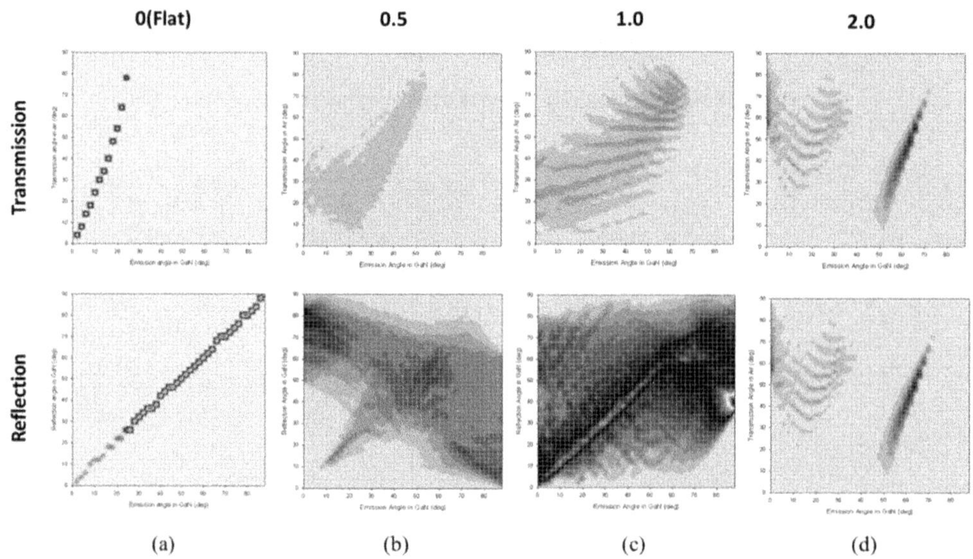

Figure 3.27. The angular and relative energy distribution of the transmission and the reflection with a microlens of which the aspect ratio is (a) 0, (b) 0.5, (c) 1.0, and (d) 2.0.

However, if the energy of the light that can be transmitted is integrated, the total energy of transmission does not seem to increase compared to a surface without any microstructure. In other words, the microstructure seems unable to enlarge the LEC as you expected.

The microstructure does cause light scattering, but it cannot directly increase the proportion of forward scattering, that is, the size of LEC. So, how does the microstructure improve LEE? The answer is that backward scattering light also changes the light path, thereby increasing the probability of the light re-entering the LEC range. In other words, if the light cannot be extracted for the first time after the backward scattering light changes direction, there is still a chance to extract the light next time. This is the concept of cavity photon recycling as mentioned in section 3.4. The cavity photon recycling process is illustrated in figure 3.28. Figure 3.29 shows the accumulated LEE of the two microstructures under different cavity photon recycling times. Regardless of the microstructure, the LEE with once photon recycling only accounts for a small portion of the final LEE. Therefore, by increasing the number of times of cavity photon recycling, the accumulated LEE can increase accordingly.

Furthermore, the change of the geometrical parameters of the microstructure will also affect the LEE. Optimization by The Monte Carlo ray-tracing method is the fastest and most direct way to figure out the best structure. For example, according to the simulation results of figure 3.29, we may avoid making 45° micropyramid, or find a feasible process to make the microlens aspect ratio able to reach 1.0.

The Monte Carlo ray-tracing method can handle most LEE problems. By creating a simple scattering model or solid geometric model, the influence of random or periodic microstructures on LEE can be effectively analyzed. However, when the

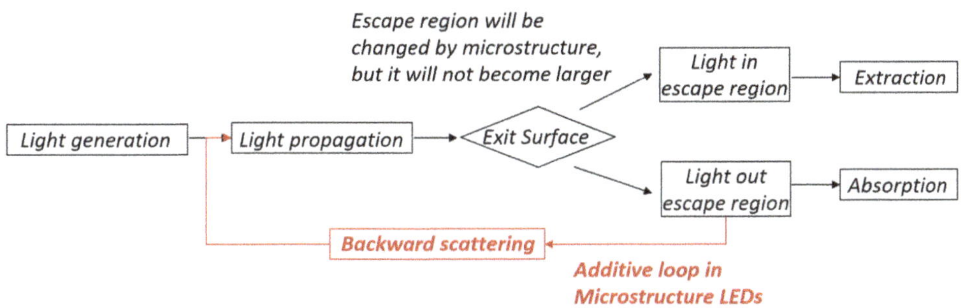

Figure 3.28. The mechanism of LED light extraction by microstructure.

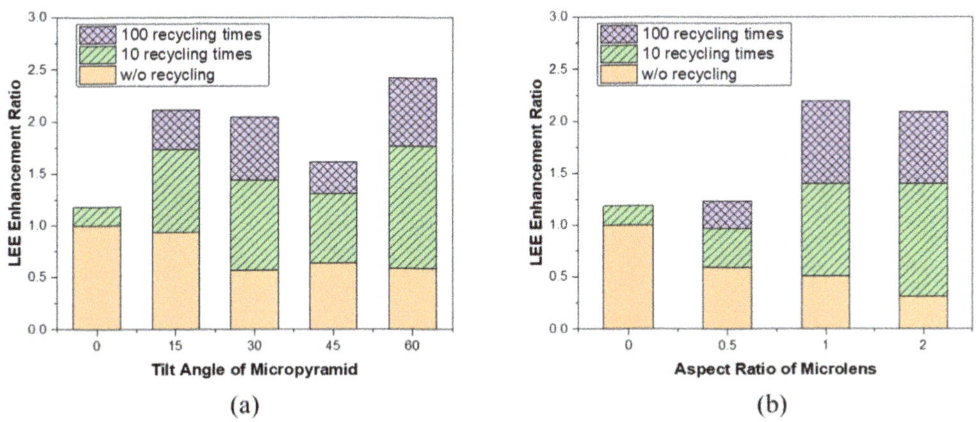

Figure 3.29. The LEE enhancement ratio by the (a) micropyramid and (b) microlens array with different recycling times.

dimension of microstructure is close to or smaller than the wavelength of light, the wave nature of light cannot be ignored. As shown in figure 3.30, the scattering effects caused by micro-scale and nano-scale microstructures are completely different. One is produced by refraction, and the other is produced by diffraction. Figure 3.31 shows the measurement of light scattering caused by blue light incident on patterned sapphire substrates with three different sizes of the microstructures. Obviously, the second and third microstructures have strong diffraction behavior, regardless of forward or backward scattering.

To simulate the diffraction behavior of light passing through microstructures, some studies choose to use numerical methods to solve electromagnetic wave problems, such as rigorous coupled wave analysis (RCWA), finite difference time domain (FDTD) and finite element method (FEM) [72–74]. Although electromagnetic wave-based methods can replace ray-based methods to simulate LEDs

Figure 3.30. The scattering effects caused by (a) refravtive and (b) diffractive microstructures.

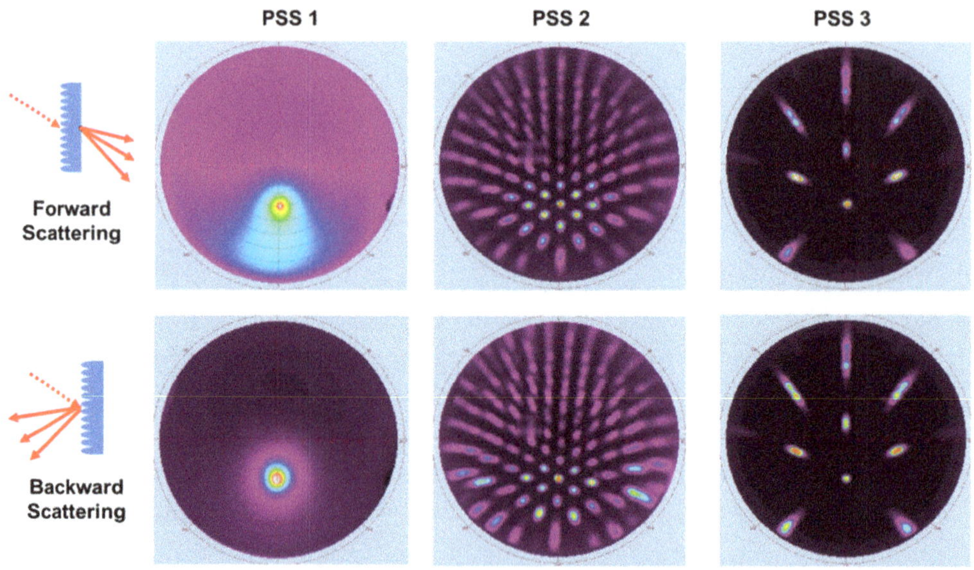

Figure 3.31. Forward and backward light scattering far-field patterns of PSS with three different scale microstructures.

with nano-scale microstructures, it still has some limitations, especially in simulating multiple interactions between the LED boundary and microstructure in the full 3D domain, such as the cavity photon recycling effect, requiring considerable computing time. In addition, to simulate the behavior of spontaneous emission, wave-based methods usually require a complicated and lengthy modeling process to simulate the random characteristics of emitted photons in the active layer.

To overcome the respective limitations of the ray-based and wave-based methods, a hybrid method combining the advantages of the two can effectively assist in the analysis of LEE. Three steps are required to combine wave and ray-tracing methods to simulate LEE. First, use electromagnetic wave simulation to calculate and collect the data sets of forward and backward light scattering of incident light at different angles after passing through the microstructure. Next, convert these data sets into an equivalent BSDF model and replace the physical microstructures in the LED model based on the Monte Carlo ray-tracing. Choosing a suitable and robust BSDF model to link wave-based and ray-based methods is an important key. Finally, perform ray tracing and calculate LEE. With this hybrid method, the problem that the ray-tracing method cannot simulate nano-scale microstructures can be overcome. Meanwhile, it also avoids the limitation that with the electromagnetic wave method it is not easy to simulate a complete LED device. For the detailed simulation process, please refer to the literature [75].

Here is shown a case study through hybrid method. The object is a GaN-on-sapphire LED with a patterned substrate, on which the microstructure is a periodic array of pyramid, whose value of size and period are equal and fixed, including 0.1, 0.5, 1.0, and 4 μm. When the emission wavelength λ is set at 450 nm, the feature size of the microstructure covers a wavelength scale ranging from 0.2 λ to 9 λ. Figure 3.32 shows the LEE simulation results of these microstructures at different aspect ratios. For the 9 λ microstructure which belongs to the micro-scale, it can be observed that

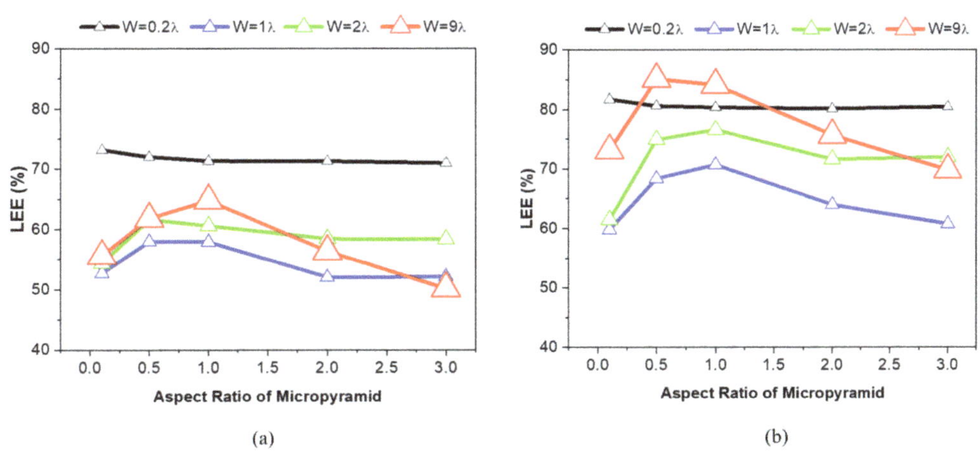

Figure 3.32. The LEE simulation results of PSS LEDs with different micropyramid structure scales under (a) bare die and (b) encapsulation using wave and ray hybrid method.

LEE has a strong dependence on changes in aspect ratio. When the aspect ratio is close to 1, LEE has the best performance, and is of the same trend as the results of the pure ray-tracing simulation. When the microstructure size is reduced to near-wavelength scales of 1 λ and 2 λ, the LEE trend is similar to that of the micro-scale microstructure, but the optimal LEE is relatively lower than that of the micro-scale microstructure, especially when the microstructure size is close to the emission wavelength. This result indicates that LEE may be suppressed when strong diffraction effects are dominant. If the microstructure size is further refined to a sub-wavelength scale of 0.2 λ, the dependence of the LEE on the aspect ratio of the microstructure becomes lower, and the LEE improves. This may be caused by the wide-angle anti-reflection property of the sub-wavelength structures [76]. The LEE trend of the packaged LED is similar to that of the bare die, and the overall LEE of all microstructures is enhanced due to changes in the external refractive index. For the case of bare die, the sub-wavelength microstructure maintains the highest LEE at all aspect ratios. However, for the packaged LED, the LEE of the micro-scale microstructure with an aspect ratio between 0.3 and 1.4 may be higher than the LEE of the sub-wavelength structure.

With the light scattering simulation of microstructure, it is proven that regardless of whether the microstructure is distributed randomly or periodically, on the surface or the interface, or at the micron- or nano-scale, these microstructures seem to be able to effectively enhance LEE. However, they may increase the probability of light extraction through different mechanisms or effects. Therefore, it is necessary to choose an appropriate simulation method before optimizing the microstructure.

References

[1] Peng W C and Wu Y C S 2006 Improved luminance intensity of InGaN–GaN light-emitting diode by roughening both the *p*-GaN surface and the undoped-GaN surface *Appl. Phys. Lett.* **89** 041116

[2] Xi J, Luo H, Pasquale A J, Kim J K and Schubert E F 2006 Enhanced light extraction in GaInN light-emitting diode with pyramid reflector *IEEE Photon. Technol. Lett.* **18** 2347–9

[3] Lee Y J, Hwang J M, Hsu T C, Hsieh M H, Jou M J, Lee B J, Lu T C, Kuo H C and Wang S C 2006 Enhancing the output power of GaN-based LEDs grown on wet-etched patterned sapphire substrates *IEEE Photonics Technol. Lett.* **18** 1152–54

[4] Lee T X, Gao K F, Chien W T and Sun C C 2007 Light extraction analysis of GaN-based light-emitting diodes with surface texture and/or patterned substrate *Opt. Express* **15** 6670–6

[5] Baba T, Inoshita K, Tanaka H, Yonekura J, Ariga M, Matsutani A, Miyamoto T, Koyama F and Iga K 1999 Strong enhancement of light extraction efficiency in GaInAsP 2-D-arranged microcolumns *J. Lightwave Technol.* **17** 2113–20

[6] Byeon K J, Cho J Y, Song J O, Lee S Y and Lee H 2013 High-brightness vertical gan-based light-emitting diodes with hexagonally close-packed micrometer array structures *IEEE Photonics J.* **5** 8

[7] Li X F, Huang S W, Lin H Y, Lu C Y, Yang S F, Sun C C and Liu C Y 2015 Fabrication of patterned sapphire substrate and effect of light emission pattern on package efficiency *Opt. Mater. Express* **5** 1784–91

[8] Chen C Y and Liu W C 2017 Light extraction enhancement of gan-based light-emitting diodes with textured sidewalls and ICP-transferred nanohemispherical backside reflector *IEEE Trans. Electron Devices* **64** 3672–7

[9] Chen W C, Niu J S, Liu I P, Wang Z F, Cheng S Y and Liu W C 2021 Study of a GaN-based light-emitting diode with a specific hybrid structure *ECS J. Solid State Sci. Technol.* **10** 7

[10] Choi H S and moon C H 2016 Improvement of the light extraction efficiency of LED package using double-layered encapsulant with curved interface *J. Nanoelectron. Optoelectron.* **11** 170–4

[11] Dang S H, Li C X, Li T B, Jia W, Han P D and Xu B S 2014 Light-extraction enhancement and directional emission control of GaN-based LED with transmission grating *Optik* **125** 3623–7

[12] Ee Y K, Kumnorkaew P, Arif R A, Tong H, Gilchrist J F and Tansu N 2009 Light extraction efficiency enhancement of InGaN quantum wells light-emitting diodes with polydimethylsiloxane concave microstructures *Opt. Express* **17** 13747–57

[13] Ee Y K, Kumnorkaew P, Arif R A, Tong H, Zhao H P, Gilchrist J F and Tansu N 2009 Optimization of light extraction efficiency of III-nitride LEDs with self-assembled colloidal-based microlenses *IEEE J. Sel. Top. Quantum Electron.* **15** 1218–25

[14] Gao X M, Li X, Yang Y C, Yuan W, Xu Y, Cai W and Wang Y J 2017 Suspended p–n junction InGaN/GaN multiple quantum wells device with bottom silver reflector *Opt. Commun.* **395** 82–7

[15] Horng R H, Huang S H, Yang C C and Wuu D S 2006 Efficiency improvement of GaN-based LEDs with ITO texturing window layers using natural lithography *IEEE J. Sel. Top. Quantum Electron.* **12** 1196–201

[16] Jiang W J, Xu C, Chen Y X, Shen G D and Song X W 2007 Increase in extraction efficiency of AlGaInP-based light-emitting diodes by surface texture *J. Cent. South Univ. Technol.* **14** 24–6

[17] Kim S J, Kim K H, Chung H Y, Shin H W, Lee B R, Jeong T, Park H J and Kim T G 2014 Light-output enhancement of GaN-based vertical light-emitting diodes using periodic and conical nanopillar structures *Opt. Lett.* **39** 3464–7

[18] Kumar P, Khanna A, Son S Y, Lee J S and Singh R K 2011 Analysis of light out-coupling from microlens array *Opt. Commun.* **284** 4279–82

[19] Lee J H, Lee D Y, Oh B W and Lee J H 2010 Comparison of InGaN-based LEDs grown on conventional sapphire and cone-shape-patterned sapphire substrate *IEEE Trans. Electron Devices* **57** 157–63

[20] Lee J H, Oh J T, Kim Y C and Lee J H 2008 Stress reduction and enhanced extraction efficiency of GaN-based LED grown on cone-shape-patterned sapphire *IEEE Photonics Technol. Lett.* **20** 1563–5

[21] Lee Y J, Hwang J M, Hsu T C, Hsieh M H, Jou M J, Lee B J, Lu T C, Kuo H C and Wang S C 2006 GaN-based LEDs with Al-deposited V-shaped sapphire facet mirror *IEEE Photonics Technol. Lett.* **18** 724–6

[22] Lee Y J, Lee C J and Chen C H 2011 Effect of surface texture and backside patterned reflector on the algainp light-emitting diode: high extraction of waveguided light *IEEE J. Quantum Electron.* **47** 636–41

[23] Lee Y L and Liu W C 2018 Enhanced light extraction of GaN-based light-emitting diodes with a hybrid structure incorporating microhole arrays and textured sidewalls *IEEE Trans. Electron Devices* **65** 3305–10

[24] Liu D S, Lin T W, Huang B W, Juang F S, Lei P H and Hu C Z 2009 Light-extraction enhancement in GaN-based light-emitting diodes using grade-refractive-index amorphous titanium oxide films with porous structures *Appl. Phys. Lett.* **94** 3

[25] Park H, Byeon K J, Yang K Y, Cho J Y and Lee H 2010 The fabrication of a patterned ZnO nanorod array for high brightness LEDs *Nanotechnology* **21** 7

[26] Chang C H, Lee Y L, Wang Z F, Liu R C, Tsai J H and Liu W C 2019 Performance improvement of GaN-based light-emitting diodes with a microhole array, 45 degrees sidewalls, and a SiO_2 Nanoparticle/Microsphere Passivation Layer *IEEE Trans. Electron Devices* **66** 505–11

[27] Li Y, Huang Y and Lai Y 2009 Investigation of efficiency droop behaviors of InGaN/GaN multiple-quantum-well LEDs with various well thicknesses *IEEE J. Sel. Top. Quantum Electron.* **15** 1128–31

[28] Miyoshi M, Tsutsumi T, Kabata T, Mori T and Egawa T 2017 Effect of well layer thickness on quantum and energy conversion efficiencies for InGaN/GaN multiple quantum well solar cells *Solid State Electron. Lett.* **129** 29–34

[29] Krames M R *et al* 1999 High-power truncated-pyramid $(Al_xGa_{1-x})0.5\ In_{0.5}P/GaP$ light-emitting diodes exhibiting >50% external quantum efficiency *Appl. Phys. Lett.* **75** 2365–7

[30] Harle V *et al* 2000 GaN-based LEDs and lasers on SiC *Phys. Status Solidi A: Appl. Res* **180** 5–13

[31] Thompson D B, Murai A, Iza M, Brinkley S, DenBaars S P, Mishra U K and Nakamura S 2008 Hexagonal truncated pyramidal light emitting diodes through wafer bonding of ZnO to GaN, laser lift-off, and photo chemical etching *Jpn. J. Appl. Phys.* **47** 3447–9

[32] Bergh A A and Saul R H 1973 Surface roughness of electroluminescent *US Patent* 3,739,217

[33] Schnitzer I and Yablonovitch E 1993 30% External quantum efficiency from surface textured, thin-film light-emitting diodes *Appl. Phys. Lett.* **63** 2174

[34] Huang H W, Chu J T, Kao C C, Hseuh T H, Lu T C, Kuo H C, Wang S C and Yu C C 2005 Enhanced light output of an InGaN/GaN light emitting diode with a nano-roughened p-GaN surface *Nanotechnology* **16** 1844–8

[35] Park S J, Sadasivam K G, Chung T H, Hong G C, Kim J B, Kim S M, Park S H, Jeon S R and Lee J K 2008 Improved light extraction efficiency in GaN-based light emitting diode by nano-scale roughening of p-GaN surface *J. Nanosci. Nanotechnol.* **8** 5393–7

[36] Wei T B, Kong Q F, Wang J X, Li J, Zeng Y P, Wang G H, Li J M, Liao Y X and Yi F T 2011 Improving light extraction of InGaN-based light emitting diodes with a roughened p-GaN surface using CsCl nano-islands *Opt. Express* **19** 1065–71

[37] Zhuo X J *et al* 2014 Improvement of light extraction efficiency in InGaN/GaN-based light-emitting diodes with a nano-roughened p-GaN surface *J. Mater. Sci.-Mater. Electron.* **25** 4200–5

[38] Tu S H, Chen J C, Hwu F S, Sheu G J, Lin F L, Kuo S Y, Chang J Y and Lee C C 2010 Characteristics of current distribution by designed electrode patterns for high power ThinGaN LED *Solid-State Electron.* **54** 1438–43

[39] Raypah M E, Devarajan M, Ahmed A A and Sulaiman F 2018 Thermal characterizations analysis of high-power ThinGaN cool-white light-emitting diodes *J. Appl. Phys.* **123** 10

[40] Raypah M E, Devarajan M and Sulaiman F 2018 Effects of solder paste on thermal and optical performance of high-power ThinGaN white LED *Solder. Surf. Mt. Technol.* **30** 182–93

[41] Raypah M E, Mahmud S, Devarajan M and AlShammari A S 2020 Research on dynamic thermal performance of high-power ThinGaN vertical light-emitting diodes with different submounts *Semicond. Sci. Technol.* **35** 11

[42] Fujii T, Gao Y, Sharma R, Hu E L, DenBaars S P and Nakamura S 2004 Increase in the extraction efficiency of GaN-based light-emitting diodes via surface roughening *Appl. Phys. Lett.* **84** 855–7

[43] Lin C F, Yang Z H, Zheng J H and Dai J H 2005 Enhanced light output in nitride-based light-emitting diodes by roughening the mesa sidewall *IEEE Photonics Technol. Lett.* **17** 2038–40

[44] Hsu S C, Lee C Y, Hwang J M, Su J Y, Wuu D S and Horng R H 2006 Enhanced light output in roughened GaN-based light-emitting diodes using electrodeless photoelectrochemical etching *IEEE Photonics Technol. Lett.* **18** 2472–4

[45] Jeong S M, Kissinger S, Kim D W, Lee S J, Kim J S, Ahn H K and Lee C R 2010 Characteristic enhancement of the blue LED chip by the growth and fabrication on patterned sapphire (0001) substrate *J. Cryst. Growth* **312** 258–62

[46] Kang D H, Jang E S, Song H, Kim D W, Kim J S, Lee I H, Kannappan S and Lee C R 2008 Growth and evaluation of GaN grown on patterned sapphire substrates *J. Korean Phys. Soc.* **52** 1895–9

[47] Lee Y J, Kuo H C, Lu T C, Su B J and Wang S C 2006 Fabrication and characterization of GaN-based LEDs grown on chemical wet-etched patterned sapphire substrates *J. Electrochem. Soc.* **153** G1106–11

[48] Shan L, Wei T B, Sun Y P, Zhang Y H, Zhen A G, Xiong Z, Wei Y, Yuan G D, Wang J X and Li J M 2015 Super-aligned carbon nanotubes patterned sapphire substrate to improve quantum efficiency of InGaN/GaN light-emitting diodes *Opt. Express* **23** A957–65

[49] Su Y K, Chen J J, Lin C L, Chen S M, Li W L and Kao C C 2009 Pattern-size dependence of characteristics of nitride-based LEDs grown on patterned sapphire substrates *J. Cryst. Growth* **311** 2973–6

[50] Wang D H, Xu T H and Chen F 2017 Performance enhancement of crystalline quality and optical properties of GaN epilayer grown on patterned sapphire substrate *J. Optoelectron. Adv. Mater.* **19** 517–21

[51] Narukawa Y, Narita J, Sakamoto T, Deguchi K, Yamada T and Mukai T 2006 Ultra-high efficiency white light emitting diodes *Jpn. J. Appl. Phys. Part 2—Lett. Express Lett.* **45** L1084–6

[52] Yamada M, Mitani T, Narukawa Y, Shioji S, Niki I, Sonobe S, Deguchi K, Sano M and Mukai T 2002 InGaN-based near-ultraviolet and blue-light-emitting diodes with high external quantum efficiency using a patterned sapphire substrate and a mesh electrode *Jpn. J. Appl. Phys.* **41** L1431–3

[53] Lin H Y, Chen Y J, Chang C C, Li X F, Hsu S C and Liu C Y 2012 Pattern-coverage effect on light extraction efficiency of GaN LED on patterned-sapphire substrate *Electrochem. Solid State Lett.* **15** H72–4

[54] Park H, Byeon K J, Jang J J, Nam O and Lee H 2011 Enhancement of photo- and electro-luminescence of GaN-based LED structure grown on a nanometer-scaled patterned sapphire substrate *Microelectron. Eng.* **88** 3207–13

[55] Zhou Q B, Xu M S, Li Q X and Wang H 2017 Improved efficiency of GaN-based green LED by a nano-micro complex patterned sapphire substrate *IEEE Photonics Technol. Lett.* **29** 983–6

[56] Boroditsky M, Krauss T F, Coccioli R, Vrijen R, Bhat R and Yablonovitch E 1999 Light extraction from optically pumped light-emitting diode by thin-slab photonic crystals *Appl. Phys. Lett.* **75** 1036–8

[57] Kim D *et al* 2005 Enhanced light extraction from GaN-based light-emitting diodes with holographically generated two-dimensional photonic crystal patterns *Appl. Phys. Lett.* **87** 203508–11

[58] Joyce W B, Bachrach R Z, Dixon R W and Sealer D A 1974 Geometrical properties of random particles and the extraction of photons from electroluminescent diodes *J. Appl. Phys.* **45** 2229–53
[59] Ting D Z and McGill T C Jr 1995 Monte Carlo simulation of light-emitting diode light-extraction characteristics *Opt. Eng.* **34** 3545–53
[60] Berenger J-P 1996 Perfectly matched layer for the FDTD solution of wave-structure interaction problems *IEEE Trans. Antennas Propag.* **44** 110–7
[61] Joannopoulos J D, Meade R D and Winn J N 1995 *Photonic Crystals: Modeling the Flow of Light* (Princeton, NJ: Princeton University)
[62] Leung K-M and Liu Y 1990 Photon band structures: the plane-wave method *Phys. Rev.* B **41** 10188
[63] Padjen R, Gerard J and Marzin J 1994 Analysis of the filling pattern dependence of the photonic bandgap for two-dimensional systems *J. Mod. Opt.* **41** 295–310
[64] Yee K 1966 Numerical solution of initial boundary value problems involving Maxwell's equations in isotropic media *IEEE Trans. Antennas Propag.* **14** 302–7
[65] Muth J F, Brown J D, Johnson M A L, Yu Z, Kolbas R M, Cook J W and Schetzina J F 1999 Absorption coefficient and refractive index of GaN, AlN and AlGaN alloys *J. Nitride Semicond. Res.* **4S1** G5
[66] Kim H, Park S J and Hwang H 2002 Modeling of a GaN-based light-emitting diode for uniform current spreading *Appl. Phys. Lett.* **77** 1903–4
[67] Ebong A, Arthur S and Downey E 2003 Device and circuit modeling of GaN/InGaN light emittinh diodes for optimum current spreading *Solid-State Electron.* **47** 1817–23
[68] Yun J S, Shim J I and Shin D S 2009 Enhancing current spreading by simple electrode pattern design methodology in lateral GaN/InGaN LEDs *Electron. Lett.* **45** 703–5
[69] Porch A, Morgan D V and Perks R M 2004 Transparent current spreading layers for optoelectronic devices *J. Appl. Phys.* **96** 4211
[70] Liu N, Yi X, Wang L, Sun X, Liu L, Liu Z, Wang J and Li J 2015 Light extraction improvement of blue light-emitting diodes with a metal-distributed Bragg reflector current blocking layer *Appl. Phys.* A **118** 863–7
[71] Stover J C 1995 *Optical Scattering: Measurement and Analysis* 2nd edn (Bellingham, WA: SPIE Press)
[72] Yeh W, Fang C and Chiou Y 2013 Enhancing LED light extraction by optimizing cavity and waveguide modes in grating structures *J. Disp. Technol.* **9** 359–64
[73] Zhu P, Liu G, Zhang J and Tansu N 2013 FDTD analysis on extraction efficiency of GaN light-emitting diodes with microsphere arrays *J. Disp. Technol.* **9** 317–23
[74] Liu M, Li K, Kong F, Zhao J, Ding Q and Zhang M 2016 Electrical-optical analysis of photonic crystals GaN-based high power light emitting diodes *Opt. Quant. Electron.* **48** 274
[75] Lee T X and Chou C C 2017 Scale-dependent light scattering analysis of textured structures on LED light extraction enhancement using hybrid full-wave finite-difference time-domain and ray-tracing methods *Energies* **10** 424
[76] Yang X, Zhou S, Wang D, He J, Zhou J, Li X, Gao P and Ye J 2015 Light trapping enhancement in a thin film with 2D conformal periodic hexagonal arrays *Nanoscale Res. Lett.* **10** 284

IOP Publishing

Optical Design for LED Solid-State Lighting
A guide
Ching-Cherng Sun and Tsung-Xian Lee

Chapter 4

LED package-level primary optics

The most noticeable difference between the packaging of an integrated circuit (IC) versus an LED is that the LED packaging needs to provide additional optical functions, which are usually known as the LED primary optics. The LED devices with high-quality primary optics are crucial to the design of the subsequent secondary optics. This chapter will explore these optical topics in LED packaging.

4.1 Optical considerations in LED packaging

A complete LED light source contains an LED die and its package. The LED die is the light-emitting core, and the package is responsible for protecting the die, connecting the circuit, dissipating the heat, and ensuring the final light quality. Therefore, an ideal LED package must provide the necessary support for the electrical, thermal, and optical features of the die. The typical LED packaging process includes the steps of die attachment, electrical connection, phosphor coverage, and encapsulation. Figure 4.1 shows the required materials for an LED packaging, including the lead frame substrate, reflecting housing, die-attach adhesive, bonding wire, phosphor, transparent encapsulation material, and the lens.

The high luminous efficiency is one of the most important optical considerations in LED packaging. In general, the packaging materials, structures, and processes may impact the LEE of an LED. For example, some packaging materials absorb light, so one should avoid applying them in order to not affect the LEE. Also, the encapsulation material must be transparent, and its transmittance, refractive index, shape, and size will affect the LED efficiency [1–3]. Furthermore, the size of the LED die itself, the arrangement of the LED array, the way of die attachment, the shape of the bond wire, the type of phosphor coating, among other things, will also impact the LED efficiency [4–7]. Therefore, to realize LEDs with high luminous efficiency, it is necessary to consider all the LED packaging parameters in detail and comprehensively.

doi:10.1088/978-0-7503-2368-0ch4

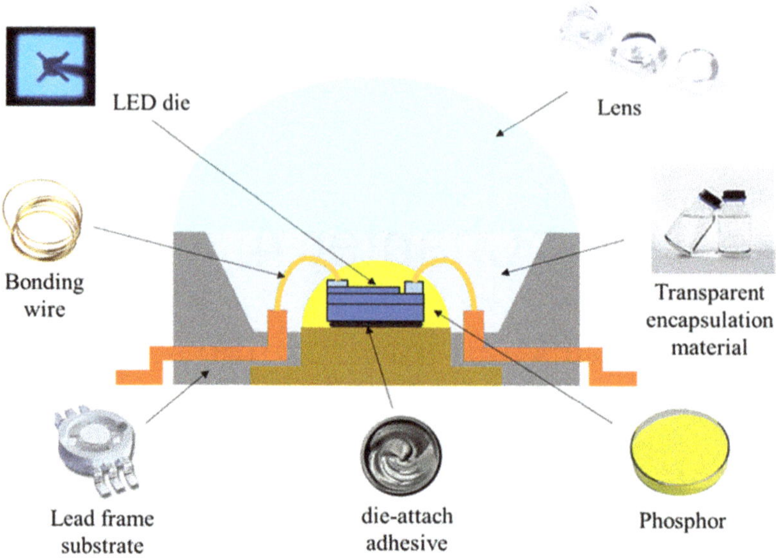

Figure 4.1. Material needed for an LED packaging.

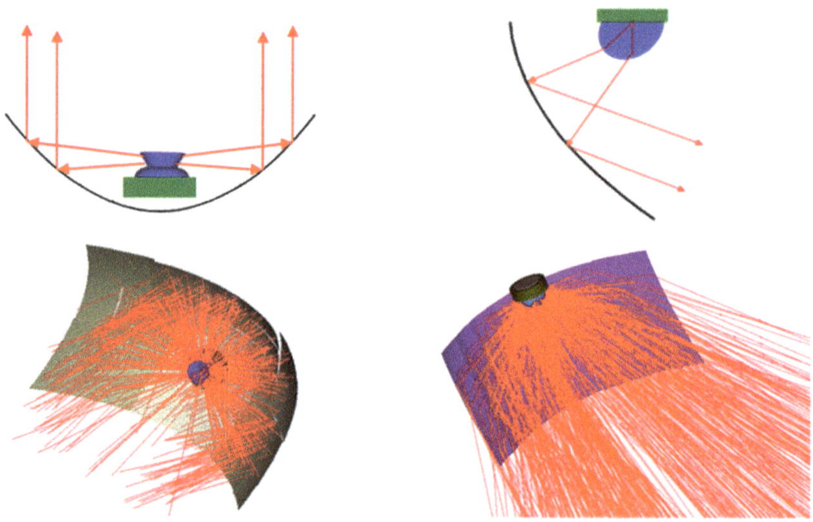

Figure 4.2. The customization of LED primary optics helps simplify secondary optics design.

Secondly, to obtain specific light distribution for various applications, the direction of the extracted light can be adjusted by shaping the encapsulation shape [8–10]. When the primary optical design can meet the required beam angle or light pattern, the secondary optical design may be simplified, by, for instance, reducing the number of optical components or minimizing the size of the optical system, as shown in figure 4.2. This approach will allow more styles to be possible in developing the optical design of an LED.

(a) (b)

Figure 4.3. White LED packaging with (a) low SCU, and (b) high SCU.

If the white-light LED packaging is to be explored further, phosphor coating or multi-color LED mixing should be considered, presuming that the white LED is to meet the requirements of general lighting [11–13]. The crucial condition is that its lighting performance must be consistent that of with the existing light source, such as the color temperature, color rendering, brightness, light distribution, and so on. Besides, the color temperature change within the viewing angle should not be easily perceived. Since the LED white light includes at least two or more color-light combinations, it is easy to produce non-uniform white light if these color lights take different paths. This feature is what's known as the spatial color uniformity (SCU). Therefore, the SCU may deteriorate when the white LED package is not well designed. For example, a yellow or blue ring may appear, as shown in figure 4.3. If these low-SCU white LEDs are further applied to the secondary optical design, the non-uniform color phenomenon will be more obvious [14]. As a result, it is essential to have high SCU performances for the high-quality white LED packaging [15–17].

Finally, for the optical system designers, the consideration of light source étendue is also necessary. A detailed description of the étendue can be found in section 6.1. In general, a light source with a smaller étendue will scale the optical system easily, especially for the optical systems that require directivity or collimation. Based on this, the LED packaging is moving towards minimizing the étendue, as shown in figure 4.4, thus giving the optical system a chance of becoming more compact [18–21].

4.2 Primary optics for high luminous efficiency

Improvements in various processes of LED packaging technology can help enhance the luminous efficiency of LED devices. Here we use ray-tracing simulation to explore the impact of three different LED packaging processes on luminous efficiency, including die mount, encapsulation, and phosphor coating.

As mentioned in section 1.4, when the LED epitaxial layer is grown on an insulating substrate, both positive and negative electrodes must be made on the epitaxial side. As a result, these LED dies with horizontal electrodes can be mounted in two ways. One is the epitaxial layer facing up and uses wire bonding for electrical connection, referred to as wire-bonding LED. Another way is to flip the LED die

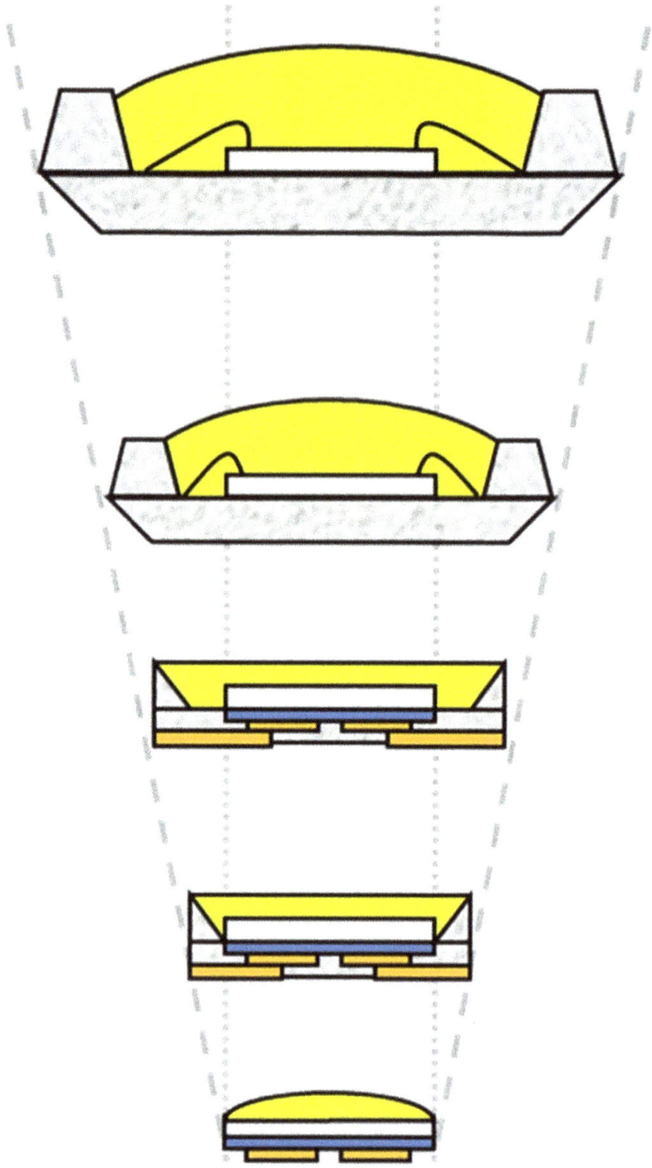

Figure 4.4. The development trend of continuous miniaturization of LED packaging.

with the epitaxial layer facing down, and the electrodes are directly bonded to the package carrier, commonly known as flip-chip LED. Figure 4.5 shows the differences between the two types. Usually, compared with wire-bonding LEDs, flip-chip LEDs have better performance in terms of heat dissipation and current spreading [22–24], so they will have a higher IQE under high-power driving. In addition, wire-bonding LEDs are generally considered to have a lower LEE and color uniformity due to the absorption of electrodes and wires [25]. In contrast, flip-chip LEDs can

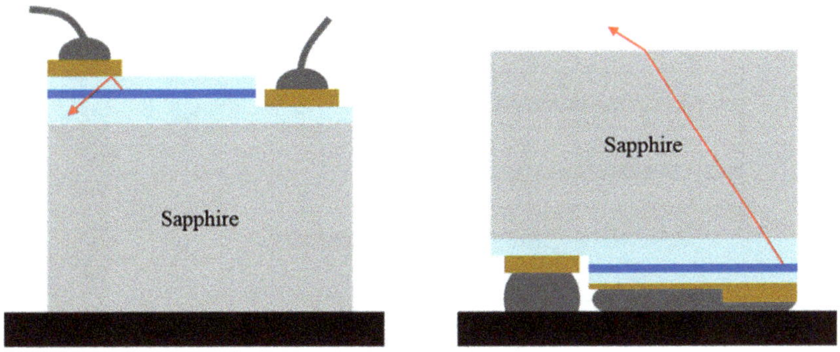

(a) (b)

Figure 4.5. The structural differences between the wire-bonding and flip-chip LED.

Table 4.1. Texturing for the different surface regions of wire-bonding and flip-chip LEDs.

Texturing / Mounting	Flat	ST	PS	ST + PS
Wire-Bonding		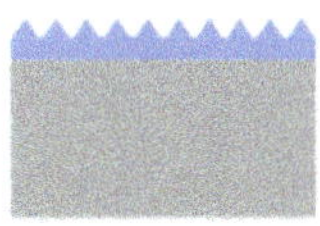		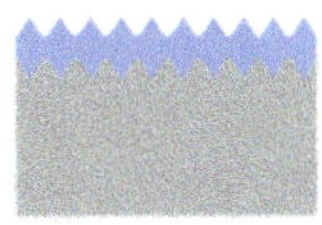
Flip-Chip		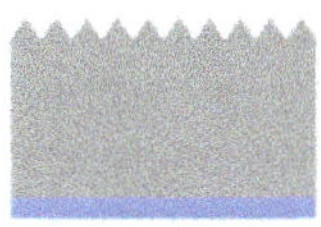	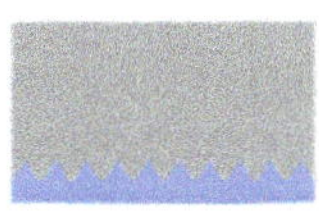	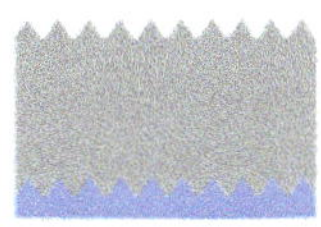

extract most light from the transparent substrate, solving the problem of electrode shielding.

Even if the flip-chip LED has excellent overall performance, the microstructure configuration of the LED die on different surface regions must be carefully considered to ensure the best LEE. Table 4.1 summarizes the various possible texturing types for wire-bonding and flip-chip GaN-on-sapphire LEDs, including surface textures (ST), patterned substrates (PS), and combinations of the two (ST+PS). It should be noted that the ST of wire-bonding is on the epitaxial layer side, while the ST of flip-chip is on the substrate side. The simulation results in figure 4.6 show that, in most cases, LEDs using flip-chip have higher LEE. However, if the LED is further encapsulated, as the refractive index difference between the sapphire (n~1.78) and the encapsulation material (n~1.50) becomes smaller, the increase in the LEE of the flip-chip LED will decrease accordingly.

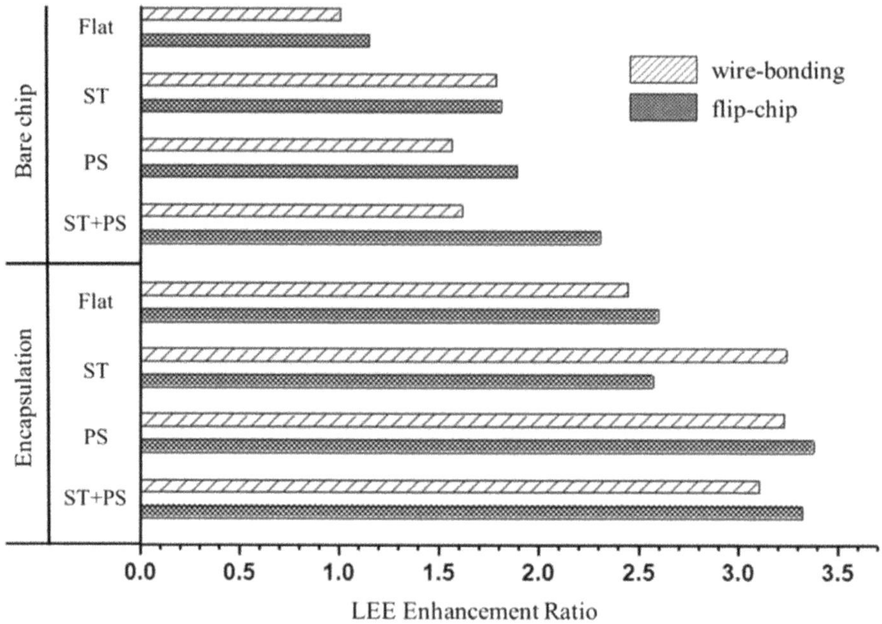

Figure 4.6. The impact of wire-bonding and flip-chip LEDs with different texturing on LEE.

For the same reason, using ST to enhance LEE does not seem suitable for flip-chip LEDs. Because in the case of encapsulation, the LEE of the flip-chip will be much lower than the wire-bonding. Therefore, if microstructures are to be used to improve LEE, PS seems to be a more suitable option for flip-chip LEDs. Another thing worth noting is that if ST and PS are applied simultaneously, LEE appears unable to increase further compared with using PS only.

In addition, according to the simulation results in figure 4.6, it can be seen that the LEE of the encapsulated LED is usually about 1.5–2.5 times higher than that of the bare die. At present, the commonly used encapsulation material for LED packaging is mainly epoxy and silicone, and their typical refractive index ranges from 1.4 to 1.65. Therefore, the refractive index of encapsulation material is just between the LED die (n: 2.4–3.5) and air (n: 1). Covering this transparent material on the LED die reduces the optical loss caused by total internal reflection. As the refractive index of the encapsulation material increases, the ratio of LEE before and after encapsulation increases accordingly. However, it is not easy to find an encapsulation material with a high refractive index and good material properties.

Besides, the encapsulation shape also affects LEE. Figure 4.7 depicts the ray path for flat and hemispherical encapsulations, respectively. Obviously, if the optimal LEE is the primary consideration, the hemispherical shape is undoubtedly the best. In this way, total internal reflection between the encapsulation and the air interface can be avoided, and the light can be effectively extracted. However, the diameter of the hemisphere must be at least 2.5 times larger than the die size. Otherwise, the light

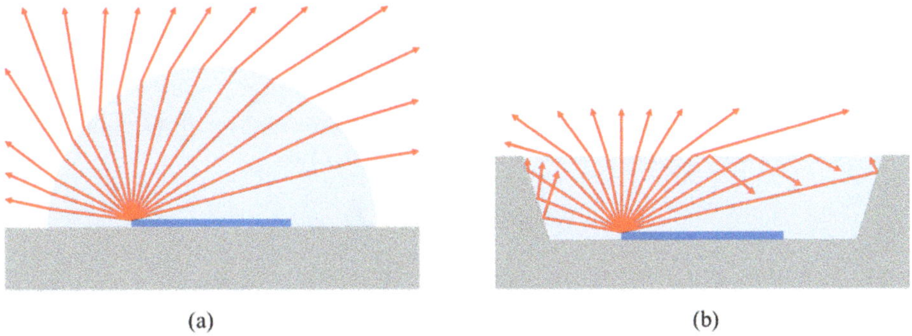

Figure 4.7. The effect of the (a) hemispherical and (b) flat top packaging geometry on the light extraction path.

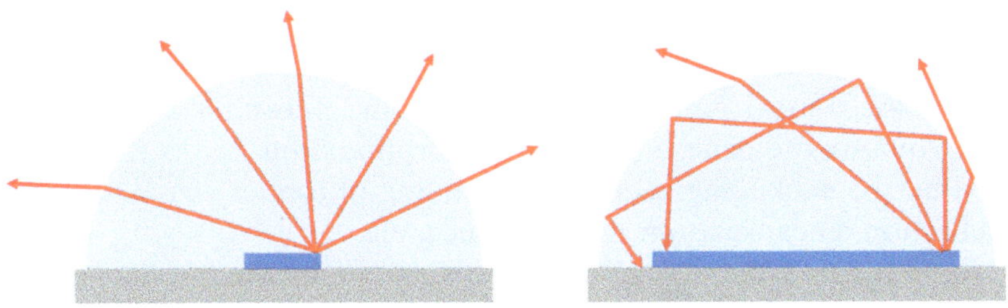

Figure 4.8. The influence of LED package and die size ratio on light extraction.

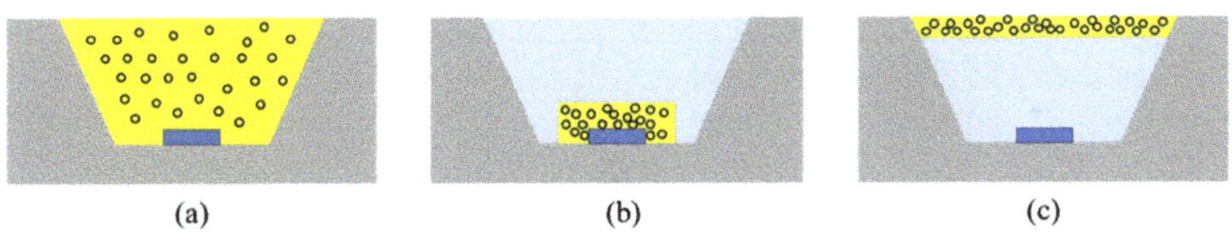

Figure 4.9. Three typical pcW-LED packaging types: (a) in-cup, (b) conformal, and (c) remote phosphor.

emitted from the edge of the LED die still has a possibility of total reflection loss, as shown in figure 4.8.

Phosphors also play a dominant role in improving the luminous efficiency of pcW-LEDs [26–31]. When the white LED needs to meet the specific requirements of color temperature and color rendering performance, it is necessary to determine the phosphor coating position, shape, concentration, thickness, etc, according to its packaging type. These phosphor packaging parameters significantly impact the luminous efficiency of white LED. Therefore, much literature has proposed many innovative designs for pcW-LED packaging to improve its white light performance, as shown in figure 4.9, including in-cup, conformal coating, and remote phosphor types. These white LED packaging issues can be effectively analyzed and optimized through phosphor optical models. The exact simulation method will be detailed in section 4.5.

The luminous efficiency of pcW-LED packages must consider optical power changes of both blue and yellow light. When the blue light is extracted from the LED die and hits the phosphor layer, part of the blue light will be absorbed and converted into yellow light; the other part is the remaining blue light. Therefore, the packaging efficiency η_{pkg} of pcW-LED is defined here as

$$\eta_{pkg} = \left(P_{b\_pkg} + P_{y\_pkg}\right)/P_{b\_die}, \tag{4.1}$$

where P_{b_die} is the optical power of blue light extracted from the LED die; P_{b_pkg} and P_{y_pkg} are the optical power of blue and yellow light extracted from the encapsulation to the ambient, respectively. In general, whether it is blue or yellow light, it may be affected by phosphor scattering and packaging geometry, resulting in a specific optical power loss. So conversely, the η_{pkg} can also be expressed as

$$\eta_{pkg} = 1 - \left(L_{pho} + L_{geo}\right)/P_{b\_die}, \tag{4.2}$$

where L_{pho} is the optical power loss caused by the phosphor itself, including the Stokes shift, quantum efficiency, and self-absorption; and L_{geo} is the optical power loss caused by the packaging geometry, including phosphor shape, die absorption, bottom reflection, encapsulation transmittance, etc.

Here, we compared the η_{pkg} of six different pcW-LEDs through the phosphor optical model. The packaging structure and simulation results are shown in figure 4.10. The first three belong to the direct cover type, and the latter three belong to the remote phosphor type. According to each phosphor shape, the

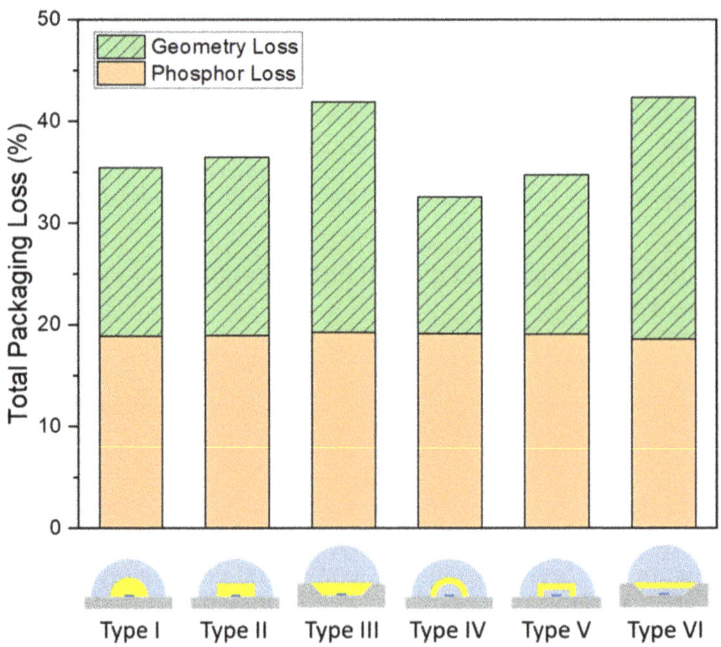

Figure 4.10. Comparison of packaging loss of six different white LED types.

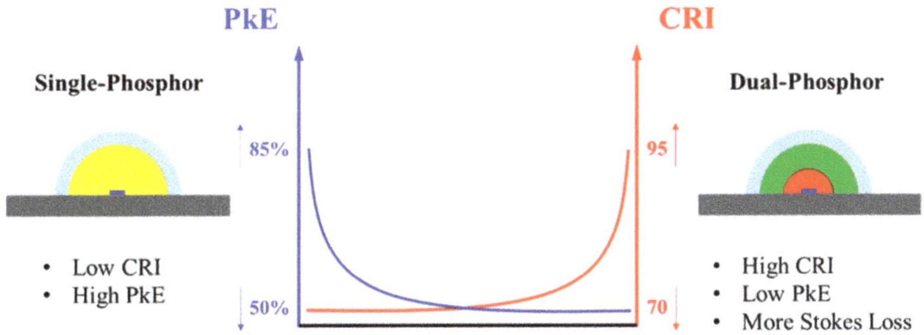

Figure 4.11. The trade-off between single-phosphor and double-phosphor LEDs in white packaging efficiency (PkE) and color rendering (CRI) performance.

phosphor concentration is adjusted to achieve the required correlated color temperature (CCT). In the case of the same CCT, the L_{pho} of pcW-LEDs is very close even in different phosphor shapes. Therefore, the overall η_{pkg} is mainly due to L_{geo}. For this reason, proposing a suitable packaging geometry and reducing material absorption will help improve the luminous efficiency of pcW-LEDs.

In general, to develop higher color rendering performance in pcW-LEDs, only a single phosphor cannot meet the CRI requirement of 80 or more. It needs to be achieved by mixing or layering phosphors with more than two colors, as shown in figure 4.11 [32, 33]. According to phosphor simulation, all the mixing phosphor types have higher η_{pkg}. Moreover, the L_{geo} of the mixing phosphor is smaller than that of the layering phosphor. The results indicate that the mixing phosphor produces weaker backward scattering, which is less affected by the reflectivity of the bottom mirror. This is mainly related to the fact that, compared with layering phosphors, mixing phosphors require less phosphor usage to achieve the target CCT. Conversely, the L_{pho} of mixing phosphors is higher than that of layering phosphors. Since both use the same phosphor, the quantum efficiency and the loss caused by the Stokes shift are the same, so the difference in L_{pho} should be the strong re-absorption effect of the mixing phosphor. For these two multi-phosphor types, although the L_{pho} is different under the same CCT, the difference is not as massive as the L_{geo}. Overall, the luminous efficiency of mixing phosphor is relatively high, and the amount of phosphor used is saved.

4.3 Primary optics to shape light distribution

The LED encapsulation not only provides necessary protection and improves LEE for the LED die, but also provides an opportunity to control its spatial light distribution (SLD). Generally, the SLD of a typical LED die is Lambertian, which can be further adjusted through the design of the encapsulation shape.

The early bullet-type LED was given this title because of the encapsulation lens shape. This type of LED is placed near the focal point of the lens and combined with the reflector cup to change the SLD, making it highly directive—that is, a narrow beam. In this way, the normal direction will get relatively high light intensity,

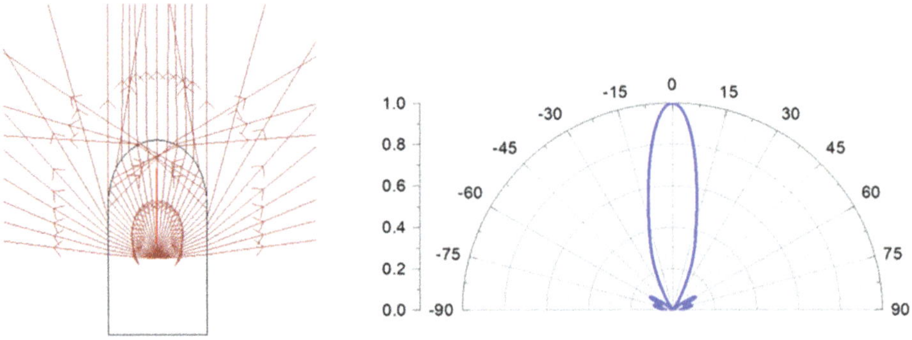

Figure 4.12. The light path and SLD of bullet-type LED.

allowing the light to be more concentrated, usually used for indication applications (figure 4.12).

Sometimes LEDs are named according to their SLD. The most famous ones are Lambertian, Batwing, and Side-emitting LEDs released by Lumileds. Lambertian LED, as the name describes, has the Lambertian distribution. It has a hemispherical lens shape, and the LED die is placed in the center of the hemisphere. In this way, the hemispherical shape can minimize the deflection of light by refraction, thereby maintaining the existing Lambertian distribution of the LED die. Because of this, it can extract light to the maximum, thus having a higher luminous efficiency. It is now a widespread LED packaging type. Batwing LED, uses a flatter (i.e. smaller curvature) curved surface design to diverge light away from the centerline to suppress the central light intensity and obtain a wider SLD, also called the batwing light pattern. According to the inverse square law of illuminance, such an SLD can help achieve flat-field illumination, uniformly illuminate the working surface, and generate no central hot spots. Side-emitting LED, uses the hollow inverted cone shape on the upper surface of the encapsulation to significantly divert the forward light to the sideward light by total internal reflection, thereby considerably reducing the forward propagation beam. Such SLD is quite suitable for designing with a reflector. Moreover, some LED candle lights are also designed according to this concept (figure 4.13).

Figure 4.14 shows the SLD changes under the size ratio between encapsulation and light-emission body (including LED die and phosphor if pcW-LEDs). It is evident from the simulation results that when the two sizes are closer, the SLD gradually becomes distorted, especially for more complex side-emitting LED. Furthermore, the refractive index of the encapsulation material will also affect the SLD, but this will not be as significant as the size effect. According to the simulation results shown in figure 4.15, the batwing LED, which mainly uses refraction to bend light, has a more noticeable change when the refractive index changes.

The optical design of encapsulation shape can make the LED light source have an SLD that meets the application needs directly. However, it is worth noting that the more complex the shape of encapsulation, the greater the risk of packaging loss.

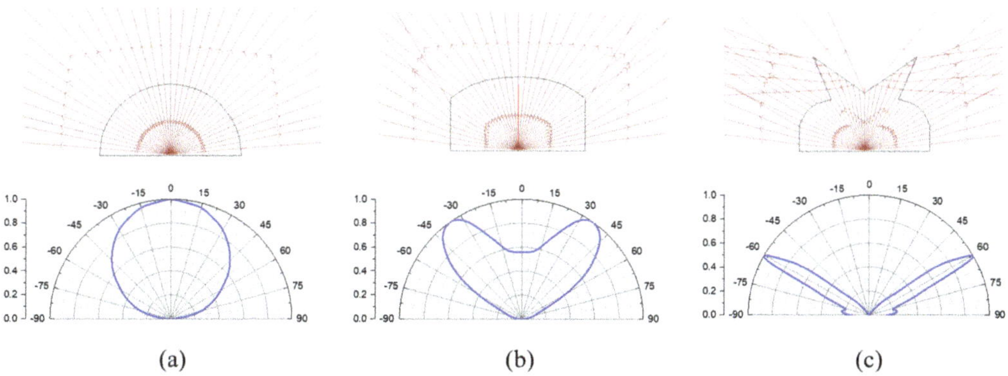

Figure 4.13. Three types of encapsulating lenses and their SLD: (a) Lambertian, (b) Batwing, and (c) side-emitting LED.

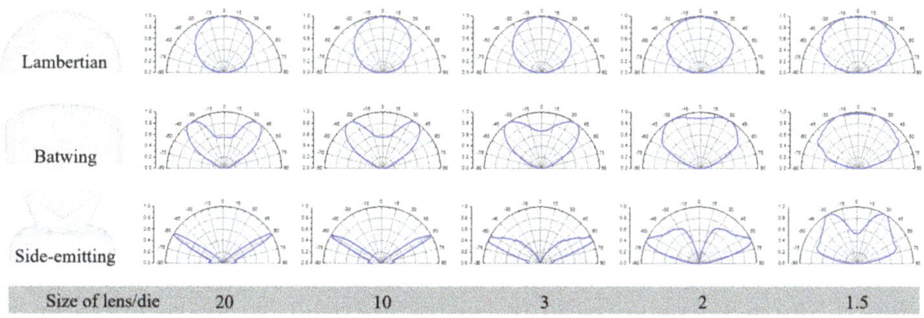

Figure 4.14. The SLD changes of the three encapsulating lenses when the lens refractive index is 1.5 under different size ratios.

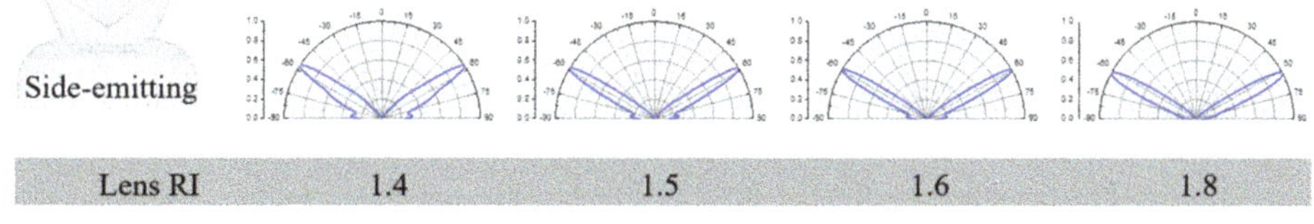

Figure 4.15. The SLD changes of the three encapsulating lenses when the size ratio is 20 under different lens refractive indexes (RI).

Therefore, one must be reminded that the LED primary optics is always the basis of the secondary optics. As long as the primary optics can provide an LED with exceptional light quality, the secondary optical design is relatively easy to achieve. Thus, in most cases, the conservative way is to let primary optics extract as much light as possible from the LED die and keep being Lambertian, and then apply the secondary optical design to meet its SLD requirements.

4.4 Primary optics for low spatial color deviation

For pcW-LEDs, different packaging types and parameters will affect the luminous efficiency and SLD, and spatial color uniformity (SCU) is another essential quality index. Such a requirement is to prevent the CCT deviation of the white LED from causing distress to users. We know that the white light generated by pc-WLEDs includes blue light emitted from the LED die and yellow light after phosphor conversion. If the SLD of each color light is slightly different, it is easy to have a white light CCT shift at different viewing angles. Therefore, SCU control is vital for white LED lighting.

At present, the commonly used phosphor coating technique is to mix the phosphor powder with silicone glue, then apply it by dispensing, spraying, or molding, and finally heating it into shape. These direct deposition methods are generally incapable of achieving a specific phosphor shape, and the phosphor itself is prone to precipitation. Thus, the distribution of the phosphor in silicone glue is challenging to control uniformly, which results in a low SCU. Such a situation may cause white LEDs to be disqualified in general lighting applications. The worst condition is that the blue leakage may be harmful to the human eye.

Figure 4.16 shows the angular CCT distribution measurement of three commercial pcW-LEDs with an average CCT around 6500 K. The angular CCT deviations

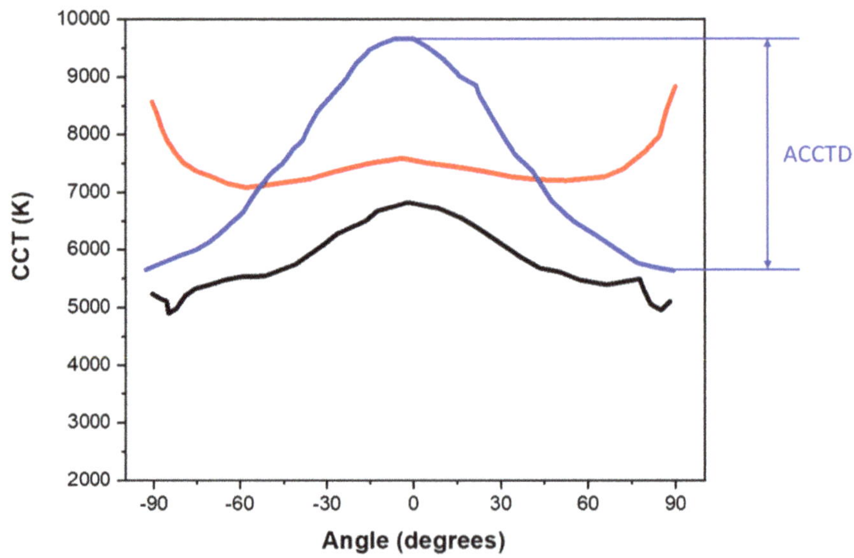

Figure 4.16. The angular CCT distribution measurement of three commercial pcW-LEDs.

(ACCTD), defined as the difference between the maximum and minimum angular CCT, are as high as 1500–3000 K. This difference is that blue and yellow light have different luminous areas and SLD due to different packaging types, resulting in a large ACCTD. However, under the same packaging type, pcW-LEDs with low CCT usually have more uniform light mixing due to the strong scattering of heavy-doped phosphors, so a low ACCTD can be obtained more easily. Therefore, in contrast, pcW-LEDs with a CCT above 5000 K need to pay more attention to the issue of high ACCTD.

A fundamental concept to have a smaller ACCTD is to enable all lights emitted from the LED die to go through an equal path in the phosphor volume. Thus choosing a light source with a spatial coherence as small as possible may be beneficial. The first idea is that the LED die can emit collimating light, so that the phosphor can be made a planar plate or film to cause each light to have an equal path in the phosphor. However, even though we have a collimating light source, due to the scattering property of phosphor, the SLD may be different after the blue and yellow light are extracted. The second idea is that the LED die selected be as small as possible to be regarded as a point source. Then, the phosphor can be formed into a hemisphere to achieve an equal path. Unfortunately, LED dies cannot be considered point light sources, especially pcW-LEDs requiring a high-power operation. In addition, ACCTD will also depend on the reflective cup and the encapsulation lens [34–38]. Therefore, to achieve an extremely small ACCTD, the geometry and configuration of the materials such as the LED die, phosphor, lens, and cup should all be designed carefully. Figure 4.17 shows the angular CCT distributions of the six packaging structures discussed in section 4.2. The results show that the pcW-LED packaging does affect ACCTD significantly.

In fact, ACCTD is mainly related to the number of collisions between blue light and phosphor particles. The number of collisions (N_{coll}) is calculated as follows:

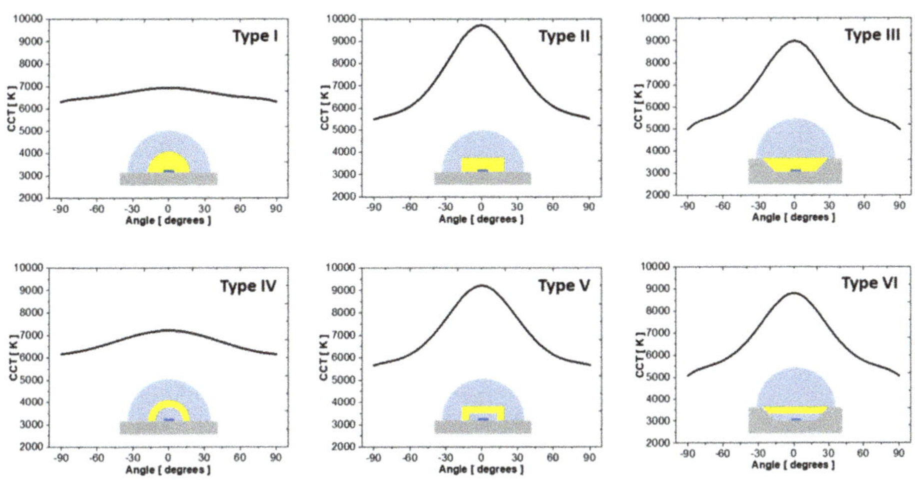

Figure 4.17. The angular CCT distributions of the six packaging structures.

$$N_{\mathrm{coll}} = n \times \sigma \times D, \tag{4.3}$$

where n is the number of phosphor particles per unit volume, σ is the effective cross-sectional area for collision, and D is the thickness of the phosphor layer. Among them, the first two products are called fraction obscuration. The reciprocal fraction obscuration represents the average light path length before colliding with the particle, also called the mean free path. Details will be explained in section 4.5. Figure 4.18 shows the inverse relationship between ACCTD and N_{coll}. Therefore, the more the blue light collides with phosphor, the more uniformly the light is mixed, and the smaller the ACCTD.

Another problem causing large ACCTD comes from the size of the LED die. In principle, if the phosphor coverage is extensive enough, the influence of die size can be reduced. However, when the phosphor size is close to the encapsulation lens, the total internal reflection caused by the lens may change the SLD of blue and yellow light again. To minimize the ACCTD of pcW-LEDs without increasing the phosphor size, other packaging parameters of the phosphor can be tried to further optimize. Taking the type I pcW-LED package in figure 4.17 as an example, a simple solution is to extend the phosphor's vertical thickness to increase the collision numbers of blue light with the phosphor along the normal direction. Figure 4.19 shows the simulated angular CCT distribution and its corresponding ACCTD change under different extended phosphor thicknesses. It can be seen from the figure that an extended thickness of about 0.3 mm can significantly reduce ACCTD. Thus the best phosphor dome is not a hemisphere. For other pcW-LED packaging types, of course, there are opportunities to achieve the smallest ACCTD through optimization.

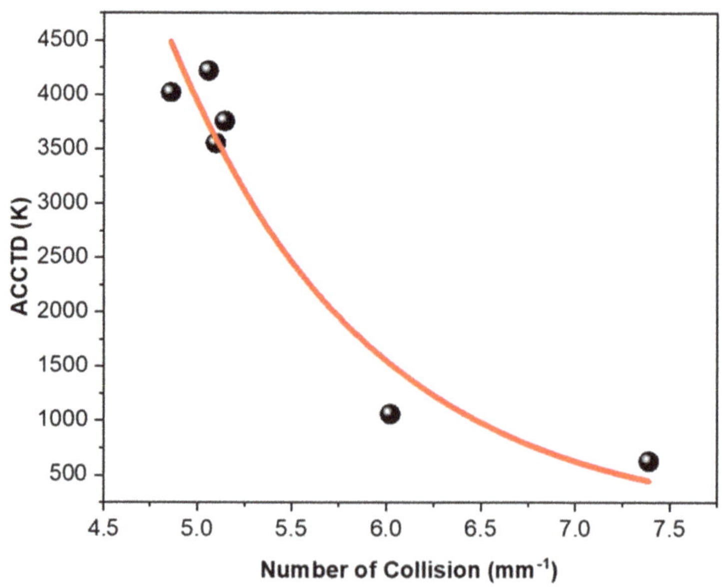

Figure 4.18. The relationship between ACCTD and collision number.

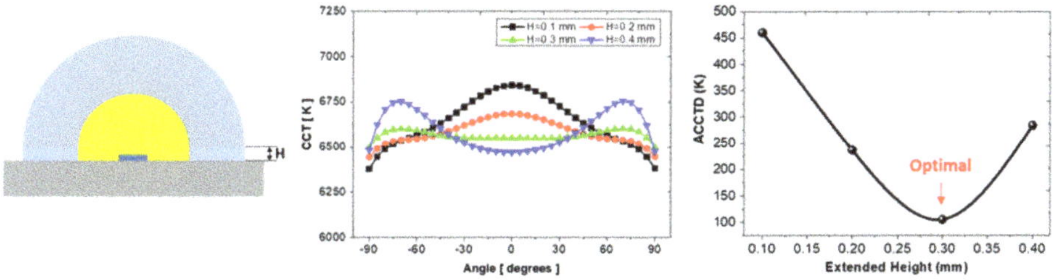

Figure 4.19. The simulated angular CCT distribution of the type I package LED under different extending phosphor thicknesses.

4.5 Phosphor modelling

To have a more thorough understanding of the optical and color performance of pcW-LEDs, it is necessary to develop an equivalent optical model for phosphor. With the help of a successful phosphor model, it is possible to quickly analyze the phosphor technology on the LED package and then obtain the packaging parameters that are ideal to white LED. It can even be further used to design various LED package types to satisfy multiple needs, so it can effectively abandon the trial and error learning method that consumes a lot of time.

Many works of literature have proposed approaches to successfully establish phosphor models [39–43]. Most of them use Monte Carlo ray-tracing combined with Mie scattering. The simulation process of the phosphor model is usually divided into two ray-tracings. The first tracing is for the short-wavelength light, which simulates the excitation light. When the excitation light enters the volume range of the simulated phosphor, the light will be scattered and absorbed by the phosphor. The absorbed energy distribution will be recorded. Then, the second tracing will re-emit long-wavelength emission light based on this absorption record. Its light intensity distribution is isotropic, and it also has a volume scattering effect in the phosphor but with no absorption. Its energy is set as the energy absorbed by excitation light multiplied by the phosphor conversion efficiency. Finally, the mixed white light is the sum of the excitation and emission light obtained from the two ray-tracings. Figure 4.20 shows a flowchart of the simulation process of a single phosphor optical model.

In general, the volume scattering of light can be divided into Rayleigh scattering and Mie scattering according to the particle size [44]. The scattering caused by the particle size less than one-tenth of the light wavelength is Rayleigh scattering, whereas Mie scattering refers to the scenario when the particle size is close to or larger than the light wavelength. The phosphors used in LEDs usually belong to Mie scattering, and the scattering distribution is related to the material, size, and quantity of the phosphor particles. A key parameter for modeling Mie scattering is fraction obscuration, which is defined as the cross-sectional area of a particle multiplied by the number of particles per unit volume, which estimates the probability of a photon hitting a particle. The reciprocal of the fraction obscuration is known as the 'mean free path,' representing the average spacing between particles.

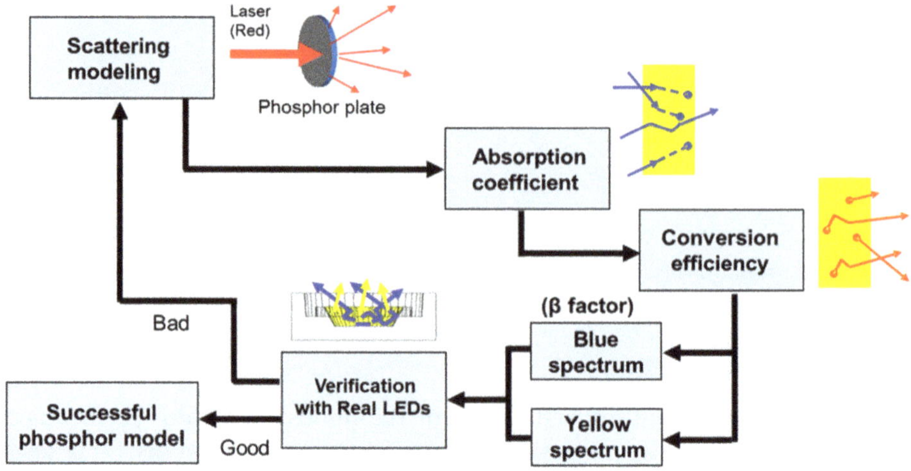

Figure 4.20. The simulation process flowchart of a single phosphor optical model. Reprinted with permission from [53].

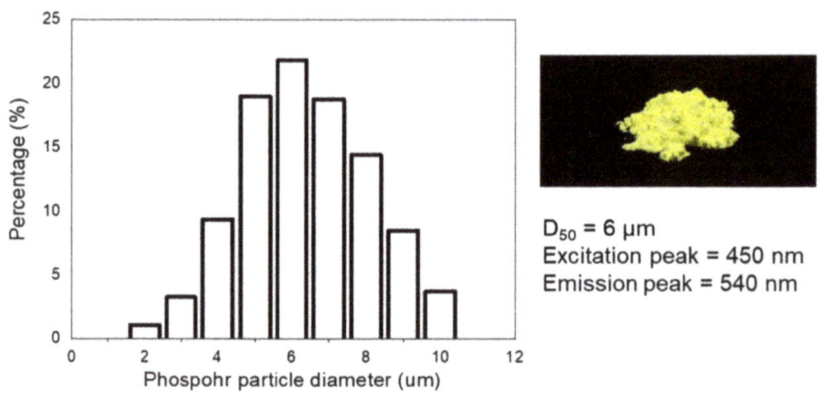

Figure 4.21. The measurement of phosphor particle size distribution.

Figure 4.21 shows the particle size distribution of a YAG:Ce phosphor measured by a particle size analyzer. The particle size range is 2–10 μm, and the median particle size D_{50} is 6 μm. To simplify the phosphor model, the phosphor in the simulation can be set to an equivalent single-particle size, which still effectively describes the scattering behavior of the phosphor. Figure 4.22 compares the scattering distribution of phosphors with single-particle size and multiple particle size distribution. The only difference is that the scattering disturbance of multiple particle size is relatively small. This difference does not significantly affect the phosphor simulation results.

The number of particles of phosphor can be estimated based on the weight percentage concentration ($w\%$):

$$w\% = \frac{W_p}{W_p + W_g}, \tag{4.4}$$

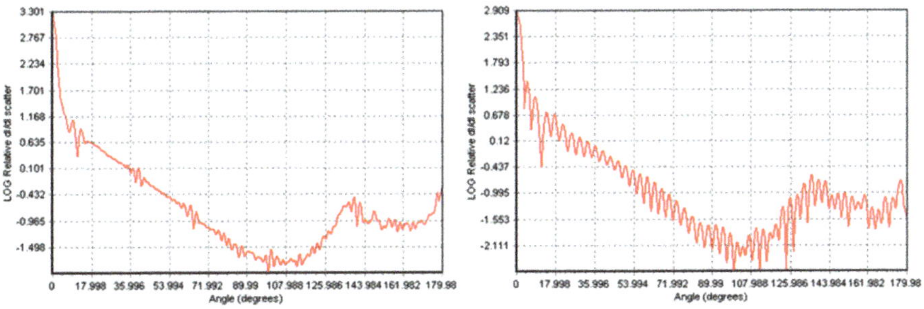

(a) (b)

Figure 4.22. The scattering intensity distribution of (a) multi-particle size and (b) single-particle size (log scale).

where W_p and W_g are the weight of phosphor and glue (silicone), respectively. It can then be converted into volume percentage concentration ($v\%$) by the respective material density, i.e.

$$v\% = \frac{\dfrac{W_p}{D_p}}{\dfrac{W_p}{D_p} + \dfrac{W_g}{D_g}}, \tag{4.5}$$

In this example, the densities of phosphor (D_p) and glue (D_g) are 4.3 and 1.0 g cm^{-3}, respectively. Then, divide the $v\%$ by the volume of a single phosphor particle to calculate the particle number per unit volume N:

$$N = \frac{v\%}{\dfrac{4}{3}\pi R^3}, \tag{4.6}$$

where R in the formula is the particle radius, from which the fraction obscuration F of the phosphor can be deduced as

$$F = N\pi R^2 = \frac{3v\%}{4R}. \tag{4.7}$$

To confirm the scattering behavior of phosphors, the direct method is to make phosphor-glue mixed plates of different thicknesses and concentrations for scattering experimental analysis, as shown in figure 4.23. The wavelength of the light source used in the experiment cannot be in the excitation spectrum of phosphor; otherwise it will cause phosphor excitation. Figure 4.24(a) shows the experimental architecture. The 632.5 nm laser light passes through a spatial filter to filter out high-frequency noise, and a confocal system is then used to obtain a more collimated light. Next, give an iris to limit the spot size (~2 mm), take the phosphor plate as the rotation center, and rotate the power meter to measure its scattering distribution. The simulation is also modeled with the same architecture, as shown in figure 4.24(b). Figure 4.25 shows some comparison results between experiments

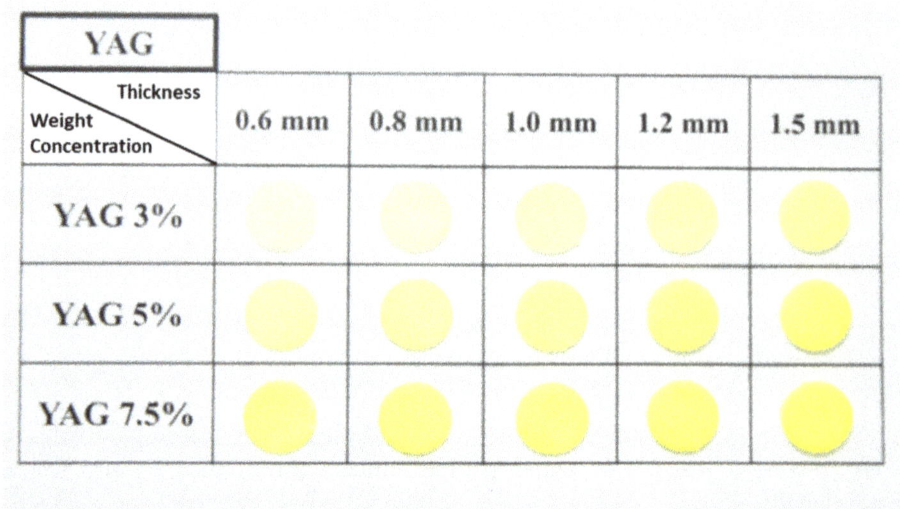

Figure 4.23. Phosphor-glue mixed plates of different thicknesses and concentrations for experimental analysis. Reprinted with permission from [54].

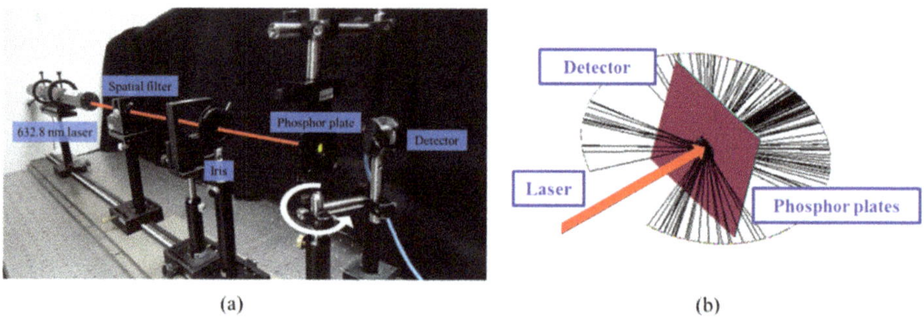

Figure 4.24. (a) Experimental setup and (b) corresponding simulation of building the phosphor scattering model. Reprinted with permission from [54, 55].

and simulations. It can be seen that both can matched. This proves that the Mie scattering model with equivalent single-particle size can be effectively applied to phosphor scattering simulation.

Another key to successfully constructing a phosphor model is to confirm its absorption coefficient and conversion efficiency. These parameters can be obtained by fitting the results from simulations to experimental data. First, it would require a standard excitation light source, as shown in figure 4.26. To obtain a more accurate phosphor model, it is recommended to select the same LED die as the real application. Otherwise it should at least have the same emission spectrum. The experiment here uses, a blue LED die EZ-700 produced by Cree, and it is mounted in a cavity of which the inner diameter is 6 mm and the height is 2 mm, and it is coated with black matt paint. The purpose is to reduce the experimental error caused by multiple reflections of light internally.

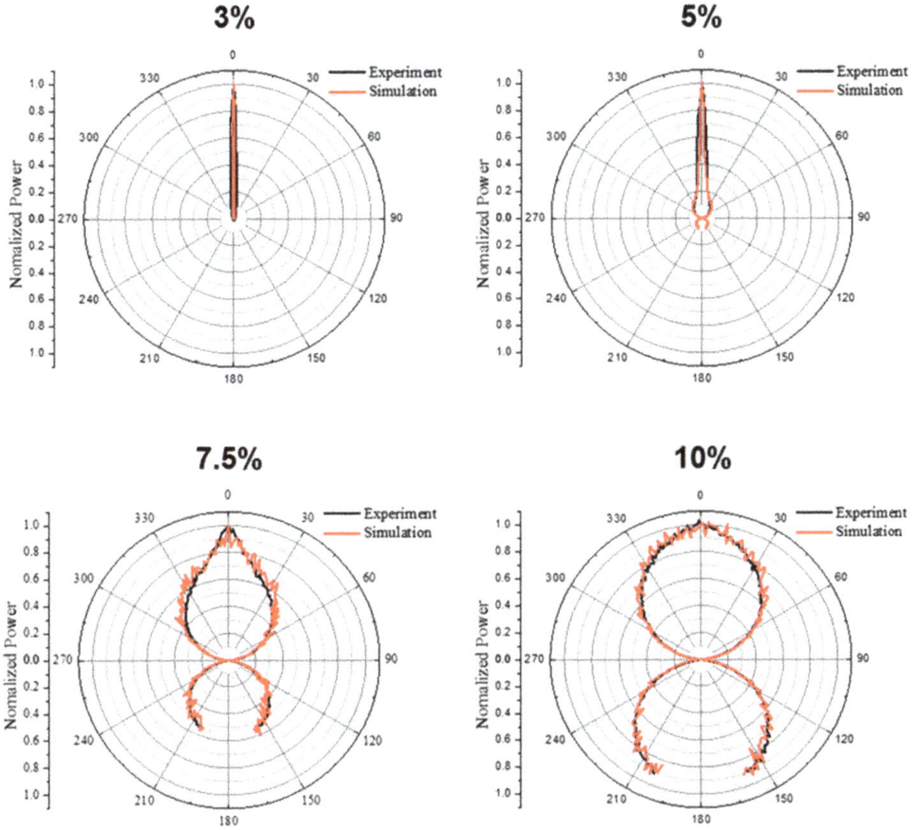

Figure 4.25. Comparison of experimental and simulated scattering of phosphor plates with different concentrations (plate thickness is 1.0 mm, concentration is 3%–10%). Reprinted with permission from [54].

The first ray tracing of the phosphor model mainly simulates the scattering and absorption of the phosphor to the excitation light source. In typical cases, the phosphor-glue mixed plate can be regarded as a uniform absorption medium. When the light is scattered, the energy will be absorbed by the phosphor simultaneously. The energy decay follows the Bouguer–Lambert–Beer Law,

$$E = E_0 e^{-\alpha L}, \tag{4.8}$$

where E_0 and E are light energy before and after passing through the phosphor, respectively. α is the absorption coefficient of the phosphor, and L is the path length of the light traveling through the phosphor. Here, the scattering of phosphor is the key to the degree of absorption. In addition, during the phosphor excitation process, affected by the Stokes shift and the quantum efficiency of the phosphor itself, only part of the blue energy absorbed by the phosphor will be re-emitted as emission light, which can be expressed as

$$E_{em} = E_{ex} \times \eta_{con}, \tag{4.9}$$

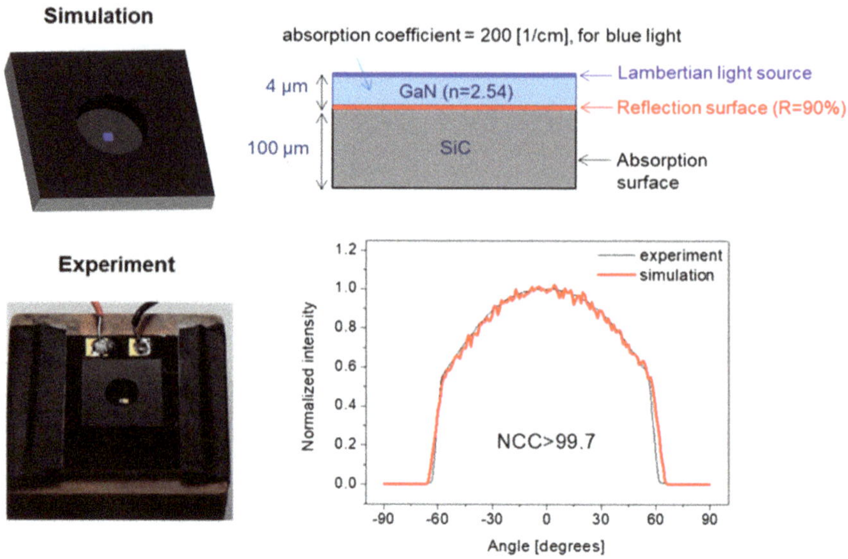

Figure 4.26. The standard excitation blue light source. Reprinted with permission from [55, 56].

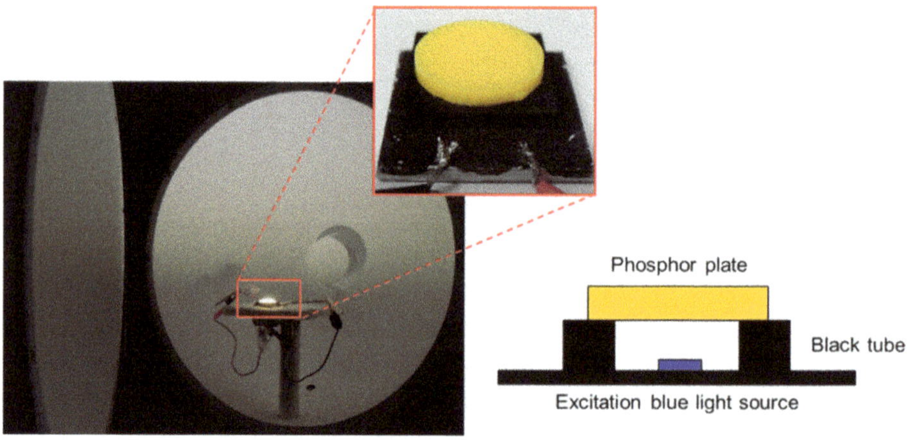

Figure 4.27. Experimental setup for measuring phosphor absorption coefficient and conversion efficiency. Reprinted with permission from [57].

where E_{em} and E_{ex} are emission light and excitation light energy, respectively. η_{con} is the conversion efficiency of the phosphor.

Figure 4.27 shows an integrating sphere system used to measure standard light sources to excite phosphors with different concentrations and thicknesses. To avoid the instability of the standard light source itself caused by thermal effects, it is recommended to drive it with a low current density. Figure 4.28 is an example of the spectrum measurement result of yellow phosphor excited by a blue LED. The spectrum can use phosphor emission spectra to separate blue and yellow light, and then calculate their respective optical power. Finally, the phosphor absorption coefficient and conversion efficiency can be determined by fitting to the experiment.

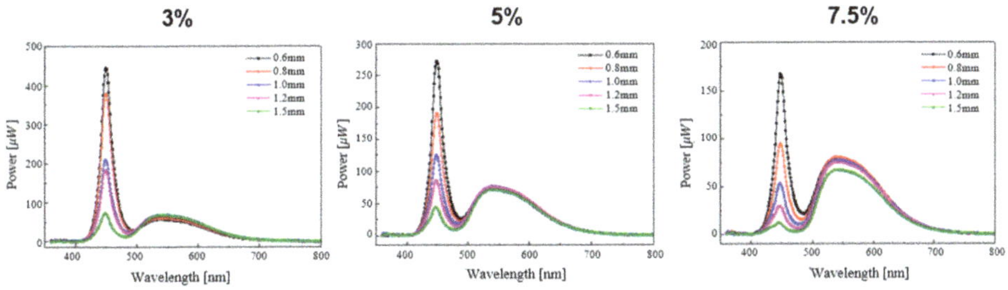

Figure 4.28. Spectra of phosphor plate with different concentration and thickness after being excited by the excitation light source.

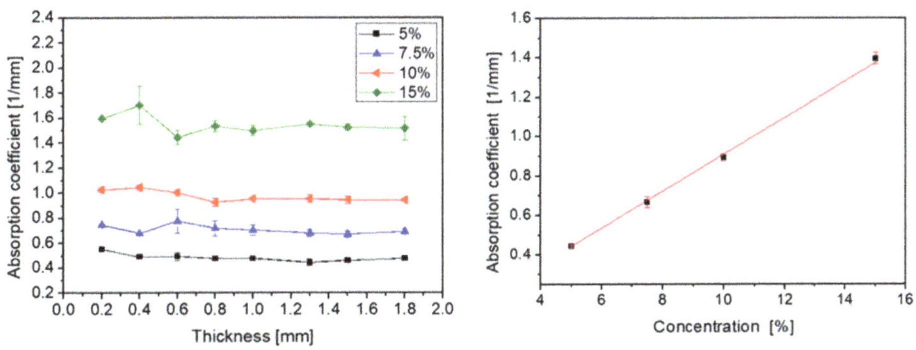

(a) (b)

Figure 4.29. (a) Calculate the absorption coefficient of blue light in phosphor. (b) Linear fitting of absorption coefficient with different phosphor concentrations.

Figure 4.29(a) shows the absorption coefficients of all tested phosphor plates at different concentrations and thicknesses. Since the absorption coefficient is defined according to Beer–Lambert's law, its value is only related to the phosphor concentration and does not change alongside the plate thickness. Therefore, the absorption coefficients of different plate thicknesses can be averaged and fitted with an equation to obtain the absorption coefficients corresponding to the concentration range between 5% and 15%, as shown in figure 4.29(b).

The absorption coefficient that was obtained is then used to simulate the absorption energy distribution of the excitation light by the phosphor. According to the measured optical power of the emission light, fitting with the simulation results can obtain another essential parameter: conversion efficiency. Based on the fluorescent principle of phosphors, the conversion efficiency of phosphors should be a particular value which does not change with the concentration and thickness of the phosphor plate, as shown in figure 4.30(a). However, in certain exceptional cases, it may be found that the conversion efficiency will decrease as the phosphor concentration and plate thickness increase, as shown in figure 4.30(b). The possible cause of this situation is the phosphor re-absorption effect. To simplify the phosphor model,

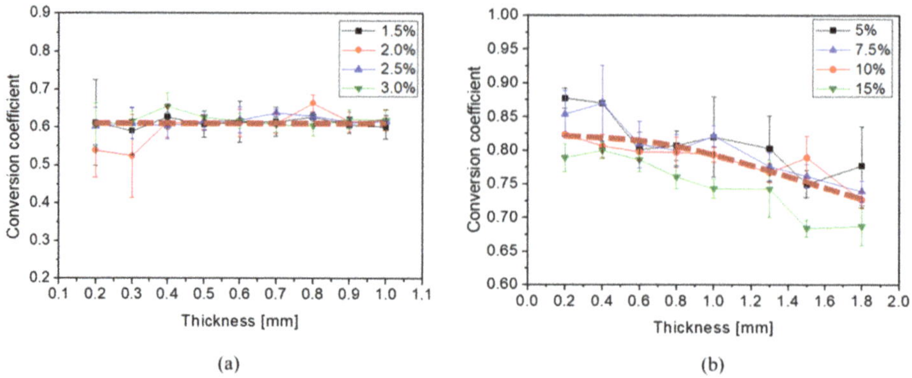

Figure 4.30. The conversion efficiency of phosphor (a) is a constant in typical case; (b) would change with the various of phosphor parameters due to the re-absorption effect.

the relationship between phosphor concentration and conversion efficiency can also be obtained through the equation fitting method.

Up until now, the simulation parameters required by the phosphor model, including Mie scattering, absorption coefficient and conversion efficiency, have been confirmed through experiments and simulations. Through the phosphor model, the optical and color performance of pcW-LEDs with different package types can be accurately predicted and analyzed. For example, the pcW-LEDs packaging efficiency, color temperature, color rendering, and spatial color deviation mentioned in chapter 4 are simulation results that were obtained using this phosphor model.

The phosphor model established above can be further extended to more complex phosphor models, such as white LEDs with multi-color phosphors, phosphors with extra scattering particles, and phosphor thermal quenching effects. For the dual-color layered phosphor model, in addition to constructing the respective phosphor models, it is also necessary to consider the spectral distortion caused by the two phosphors, and complete the modelling in four steps. Using the red and green dual-color layered phosphors excited by blue light as an example, each of the first and second steps confirms the red and green phosphor model parameters, just like the modelling method of the single phosphor model described above. However, as figure 4.31 shows, if the phosphor absorption varies with the wavelength of the excitation light, it is easy to cause distortion of the excitation light spectrum, so it must be corrected. Usually, green phosphors have this requirement, but red phosphors do not. For the detailed spectrum correction method, one can refer to the literature [45]. The third step is to build another additive model to describe the re-absorption and re-emission between the two phosphors. Here, the excitation light source is changed to the emission light of the green phosphor to confirm the absorption coefficient and conversion efficiency of the red phosphor to the green light. The Mie scattering of the two phosphors is consistent with the model established in steps 1 and 2. However, the absorption of different green light wavelengths by the red phosphor is not constant, just like the spectrum distortion of

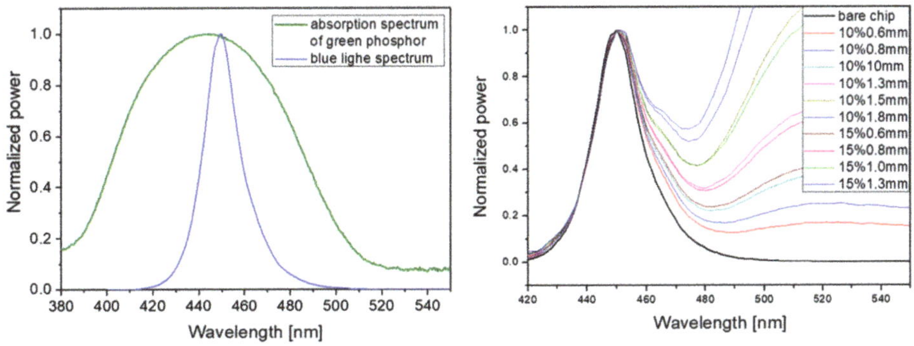

Figure 4.31. Excitation light spectrum distortion caused by blue light excited green phosphor.

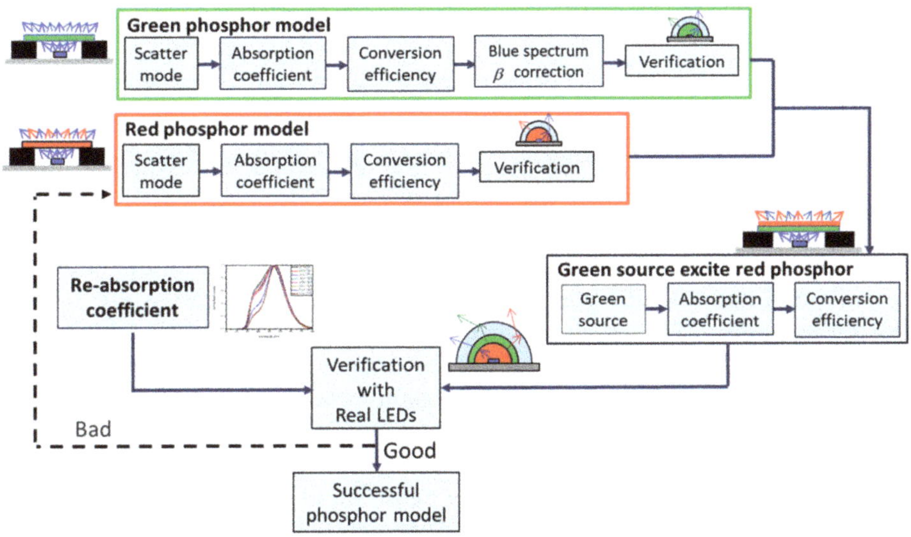

Figure 4.32. The flowchart of a dual-layered phosphor optical model.

blue light mentioned above. Therefore, the fourth step is to modify the green light spectrum and this is crucial for an accurate dual-color layered phosphor model [46]. The modelling process of the dual-color phosphor model is shown in figure 4.32.

Some references have proposed adding scattering particles to the pcW-LED package to reduce the phosphor used and increase the spatial color uniformity [47–51]. In this case, the single phosphor model can still be used as the basis. This means that the mixture of phosphor and scattering particles can be regarded as a new phosphor, and the concepts of equivalent refractive index and equivalent particle size are introduced to simulate the scattering characteristics. Similarly, the absorption coefficient and conversion efficiency can be obtained by modelling a single phosphor. The modelling process of phosphor mixed scattering particles is shown in figure 4.33.

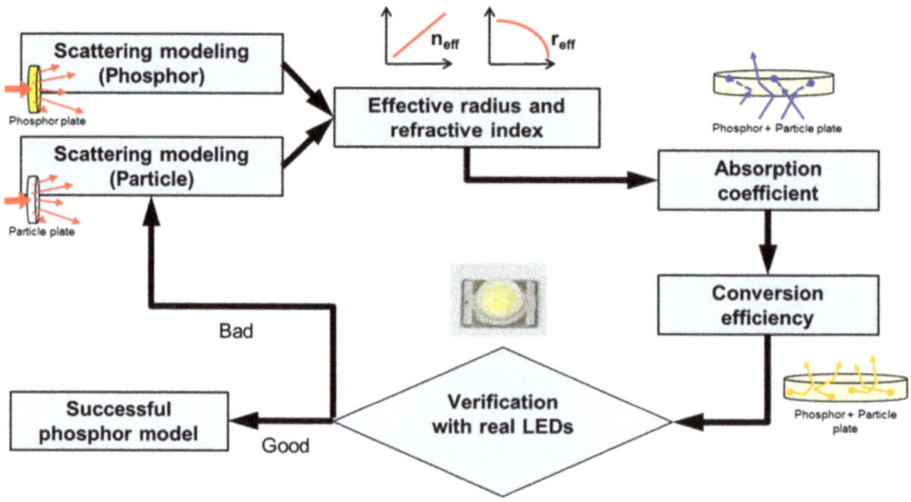

Figure 4.33. The flowchart of a phosphor mixes scattering particle optical model.

The above-mentioned phosphor models are all constructed without considering the thermal effect. However, we know that when LEDs are driven by high current density, high temperatures are unavoidable and this will cause changes in the optical and color characteristics of pcW-LEDs [52]. Therefore, if the thermal effect can be integrated into the original phosphor model, the trend of the white LED with increasing temperature can be further analyzed, and the phosphor packaging parameters can be optimized to minimize the impact on the operating temperature. In the combined optical and thermal phosphor model, the optical and thermal characteristics of the blue LED die should be simultaneously grasped. And one should be reminded that under a specific operating current for a white LED, the junction temperature of the LED die may be different from the temperature of the phosphor. In the phosphor part, the absorption coefficient and conversion efficiency of phosphor will become a function of temperature, but the temperature effect on scattering can be ignored. The absorption coefficient can be estimated through the spectral changes of the blue LED at different temperatures. Generally speaking, as the temperature gets higher, the blue LED spectrum will generate red-shift, resulting in the lower absorption coefficient. The conversion efficiency can be derived through the phosphor thermal quenching experiment. Intuitively, the higher the temperature is, the lower the conversion efficiency will be.

In the construction of the phosphor model, as short-wavelength light is absorbed and scattered in the phosphor, the long-wavelength light is emitted by the phosphor and is also scattered at the same time. Therefore, in the model, we must thoroughly grasp the complex optical characteristics of light scattering, absorption, emission and even re-absorption. This is the very reason that a lot of systematic experiments are required during the modelling process to confirm the final simulated parameters one by one so as to provide a more accurate pcW-LEDs performance estimate for the phosphor model.

References

[1] Moreno I, Bermúdez D and Avendaño-Alejo M 2010 Light-emitting diode spherical packages: an equation for the light transmission efficiency *Appl. Opt.* **49** 12–20

[2] Liu Y, Lin Z, Zhao X, Tuan C C, Moon K S, Yoo S, Jang M G and Wong C P 2014 High refractive index and transparent nanocomposites as encapsulant for high brightness LED packaging *IEEE Trans. Compon. Packag. Manuf. Technol.* **4** 1125–30

[3] Nagasawa Y and Hirano A 2019 Review of encapsulation materials for AlGaN-based deep-ultraviolet light-emitting diodes *Photonics Res* **7** B55–65

[4] Chung S C, Ho P C, Li D R, Lee T X, Yang T H and Sun C C 2015 Effect of chip spacing on light extraction for light-emitting diode array *Opt. Express* **23** A640–9

[5] Law T K, Lim F, Li Y, Puan X P, Sng G K E and Teo J W R 2017 Effect of packaging architecture on the optical and thermal performances of high-power light emitting diodes *J. Electron. Packag.* **139** 031003

[6] Park Y, Li K H, Fu W Y, Cheung Y F and Choi H W 2018 Packaging of InGaN stripe-shaped light-emitting diodes *Appl. Opt.* **57** 2452–8

[7] Alim M A, Abdullah M Z, Aziz M S A and Kamarudin R 2021 Die attachment, wire bonding, and encapsulation process in LED packaging: a review *Sens. Actuators A: Phys* **329** 112817

[8] Wang S, Wang K, Chen F and Liu S 2011 Design of primary optics for LED chip array in road lighting application *Opt. Express* **19** A716–24

[9] Zhu Z, Yan Y, Wei S, Fan Z and Ma D 2019 Compact freeform primary lens design based on extended Lambertian sources for liquid crystal display direct-backlight applications *Opt. Eng.* **58** 025108

[10] Nian L, Pei X, Zhao Z and Wang X 2019 Review of optical designs for light-emitting diode packaging *IEEE Trans. Compon. Packag. Manuf. Technol.* **9** 642–8

[11] Liu Z and Liu S 2008 Optical analysis of color distribution in white LEDs with various packaging methods *IEEE Photon. Technol. Lett.* **20** 2027–9

[12] Wu H, Narendran N, Gu Y and Bierman A 2007 Improving the performance of mixed-color white LED systems by using scattered photon extraction technique *Proc. SPIE* **6669** 666905

[13] Sun C C, Moreno I, Lo Y C, Chiu B C and Chien W T 2012 Collimating lamp with well color mixing of red/green/blue LEDs *Opt. Express* **20** A75–84

[14] Wang K, Wu D, Chen F, Liu Z, Luo X and Liu S 2010 Angular color uniformity enhancement of white light-emitting diodes integrated with freeform lenses *Opt. Lett.* **35** 1860–2

[15] Cvetkovic A, Mohedano R, Dross O, Hernandez M, Benítez P, Miñano J C, Vilaplana J and Chaves J 2012 Primary optics for efficient high-brightness LED colour mixing *Proc. SPIE* **8485** 84850Q

[16] Shuai Y, He Y Z, Tran T and Shi F G 2011 Angular CCT uniformity of phosphor converted white LEDs: effects of phosphor materials and packaging structures *IEEE Photon. Technol. Lett.* **23** 137–9

[17] Huang H T, Tsai C C and Huang Y P 2010 Conformal phosphor coating using pulsed spray to reduce color deviation of white LEDs *Opt. Express* **18** A201–6

[18] Cvetkovic A, Dross O, Chaves J, Benítez P, Miñano J C and Mohedano R 2006 Etendue-preserving mixing and projection optics for high-luminance LEDs, applied to automotive headlamps *Opt. Express* **14** 13014–20

[19] Moreno I, Rodriguez N and Basilio J C 2013 Simultaneous color-mixing and collimation within LED package *Proc. SPIE* **8841** 884102

[20] Sorgato S, Mohedano R, Chaves J, Hernández M, Cvetkovic A, Thienpont H, Benítez P, Miñano J C and Duerr F 2015 Compact étendue-preserving light-mixing optics *Opt. Express* **23** A1485–90

[21] Lee T X, Tsai M C, Chang S C and Liu K C 2016 Miniaturized LED primary optics design used for short-distance color mixing *Appl. Opt.* **55** 9067–73

[22] Wu P, Ou S, Horng R and Wuu D 2017 Improved performance and heat dissipation of flip-chip white high-voltage light emitting diodes *IEEE Trans. Device Mater. Reliab.* **17** 197–203

[23] Zhou S J, Xu H H, Liu M L, Liu X T, Zhao J, Li N and Liu S 2018 Reflector on electrical and optical properties of GaN-based flip-chip light-emitting diodes *Micromachines* **9** 650

[24] Zhou S, Liu X, Yan H, Chen Z, Liu Y and Liu S 2019 Highly efficient GaN-based high-power flip-chip light-emitting diodes *Opt. Express* **27** A669–92

[25] Wu B, Luo X, Zheng H and Liu S 2011 Effect of gold wire bonding process on angular correlated color temperature uniformity of white light-emitting diode *Opt. Express* **19** 24115–21

[26] Kim J K, Luo H, Schubert E F, Cho J, Sone C and Park Y 2005 Strongly enhanced phosphor efficiency in GaInN white light-emitting diodes using remote phosphor configuration and diffuse reflector cup *Jpn. J. Appl. Phys.* **44** L649

[27] Liu Z, Liu S, Wang K and Luo X 2009 Optical analysis of phosphor's location for high-power light-emitting diodes *IEEE Trans. Device Mater. Reliab.* **9** 65–73

[28] Lin M T, Ying S P, Lin M Y, Tai K Y, Tai S C, Liu C H, Chen J C and Sun C C 2010 Ring remote phosphor structure for phosphor-converted white LEDs *IEEE Photon. Technol. Lett.* **22** 574–6

[29] You J P, Tran N T and Shi F G 2010 Light extraction enhanced white light-emitting diodes with multi-layered phosphor configuration *Opt. Express* **18** 5055–60

[30] Hu R, Luo X, Feng H and Liu S 2012 Effect of phosphor settling on the optical performance of phosphor-converted white light-emitting diode *J. Lumin.* **132** 1252–6

[31] Kim J P and Jeon S 2016 Investigation of light extraction by refractive index of an encapsulant, a package structure, and phosphor *IEEE Trans. Compon. Packag. Manuf. Technol.* **6** 1815–9

[32] Hu R, Cao B, Zou Y, Zhu Y, Liu S and Luo X 2013 Modeling the light extraction efficiency of bi-layer phosphors in white LEDs *IEEE Photon. Technol. Lett.* **25** 1141–4

[33] Hu C, Shi Y, Feng X and Pan Y 2015 YAG:Ce/(Gd,Y)AG:Ce dual-layered composite structure ceramic phosphors designed for bright white light-emitting diodes with various CCT *Opt. Express* **23** 18243–55

[34] Sun C C, Chen C Y, Chen C C, Chiu C Y, Peng Y N, Wang Y H, Yang T H, Chung T Y and Chung C Y 2012 High uniformity in angular correlated-color-temperature distribution of white LEDs from 2800 K to 6500 K *Opt. Express* **20** 6620–30

[35] Liu Z, Liu S, Wang K and Luo X 2008 Optical analysis of color distribution in white LEDs with various packaging methods *IEEE Photon. Technol. Lett.* **20** 2027–9

[36] Shuai Y, He Y, Tran N T and Shi F G 2011 Angular CCT uniformity of phosphor converted white LEDs: effects of phosphor materials and packaging structures *IEEE Photon. Technol. Lett.* **23** 137–9

[37] Li S, Wang K, Chen F and Liu S 2012 New freeform lenses for white LEDs with high color spatial uniformity *Opt. Express* **20** 24418–28

[38] Lee T X and Chou C F 2016 Ideal luminous efficacy and color spatial uniformity of package-free LED based on a packaging phosphor-coated geometry *Appl. Opt.* **55** 7688–93

[39] Fasbender R, Li H and Winnacker A 2003 Monte Carlo modeling of storage phosphor plate readouts *Nucl. Instrum. Methods Phys. Res. A: Accel. Spectrom. Detect. Assoc. Equip.* **512** 610–8
[40] Liaparinos P F, Kandarakis I S, Cavouras D A, Delis H B and Panayiotakis G S 2006 Modeling granular phosphor screens by Monte Carlo methods *Med. Phys.* **33** 4502–7
[41] Wu H and Jenkins D R 2014 Phosphor modeling and characterization *Opt. Eng.* **53** 114107
[42] Kalyvas N and Liaparinos P 2019 Analytical and Monte Carlo comparisons on the optical transport mechanisms of powder phosphors *Opt. Mater.* **88** 396–405
[43] Zhao W, Marti J, Steinfeld A and Alwahabiab Z T 2021 Optical properties and scattering distribution of thermographic phosphors *Opt. Mater.* **122** 111741
[44] Niskanen I, Forsberg V, Zakrisson D, Reza S, Hummelgård M, Andres B, Fedorov I, Suopajärvi T, Liimatainen H and Thungström G 2019 Determination of nanoparticle size using Rayleigh approximation and Mie theory *Chem. Eng. Sci.* **201** 222–9
[45] Yang T H, Chen C Y, Chang Y Y, Glorieux B, Peng Y N, Chen H X, Chung T Y, Lee T X and Sun C C 2014 Precise simulation of spectrum for green emitting phosphors pumped by a blue LED die *IEEE Photonics J.* **6** 1–10
[46] Sun C C, Peng Y N, Chang Y Y, Chen H X and Lee T X 2021 Re-absorption effect modeling for dual-layer phosphor package of white LEDs *IEEE Photonics J.* **13** 1–10
[47] Wang P C, Su Y K, Lin C L and Huang G S 2014 Improving performance and reducing amount of phosphor required in packaging of white LEDs with TiO-doped silicone *IEEE Electron. Dev. Lett.* **35** 657–9
[48] Wang P C, Lin C L and Su Y K 2013 Enhancement of light extraction efficiency in GaN-based blue light-emitting diodes by doping TiO_2 nanoparticles in specific region of encapsulation silicone *Jpn. J. Appl. Phys.* **52** 08JG15
[49] Kang Y R, Kim K H, Kim W H, Jeon S W, Jang M S, Kwak J S and Kim J P 2013 Utilization of silicone microspheres: improving color uniformity and reducing the amount of phosphor used in white light-emitting diodes *IEEE Trans. Compon. Packag. Technol.* **3** 1453–7
[50] Chen H C, Chen K J, Lin C C, Wang C H, Han H V, Tsai H H, Kuo H T, Chien S H, Shih M H and Kuo H C 2012 Improvement in uniformity of emission by ZrO_2 nano-particles for white LEDs *Nanotechnol* **23** 265201
[51] Mont F W, Kim J K, Schubert M F, Schubert E F and Siegel R W 2008 High-refractive-index TiO_2-nanoparticle-loaded encapsulants for light-emitting diodes *J. Appl. Phys.* **103** 083120
[52] Lin C C and Liu R S 2011 Advances in phosphor for light-emitting diodes *J. Phys.Chem. Lett.* **2** 1268–77
[53] Xue Z 2017 Side-wall effect on the blue/white light pattern of sapphire-based LEDs *MSc Thesis* National Central University
[54] Hsu C H 2014 Study of YAG-phosphor modelling with adding scattering particles *MSc Thesis* National Central University
[55] Peng Y N 2012 The study of optical modelling for two phosphors with double-layer *MSc Thesis* National Central University
[56] Chen H X 2013 The study of packaging efficiency and chromatic performance of white LEDs with double-layer phosphor *MSc Thesis* National Central University
[57] Chang Y Y 2012 A study of optical modeling and evaluation of color rendering property of a dual-phosphor system *MSc Thesis* National Central University

IOP Publishing

Optical Design for LED Solid-State Lighting
A guide
Ching-Cherng Sun and Tsung-Xian Lee

Chapter 5

Light source modeling

Accurate optical models of light sources are essential to obtain an exact optical design of an illumination system. Especially for LEDs, the light source model must produce the correct light distribution before reaching the far field, since secondary LED optics are often employed very close to the LED. Moreover, a more complete LED light source model should have spatial intensity and chromaticity distribution. The improper design of factors such as LED die, phosphor, and packaging may cause deviations in the spatial color. This chapter will discuss the concepts, methods, and some case studies of LED light source modeling.

5.1 Light source characteristics

Optical design is a very important issue in solid-state lighting, which is aimed to project light into selected targets to provide appropriate illumination for a user or others [1]. The optical design technology can be divided into two main divisions. The first is imaging and the second is nonimaging [2]. As for imaging technology, a point object will be imaged onto the detector plane, and hopefully to form a point which is as small as possible. The general optical design is to use a lens or a mirror to project the wavefront reflected from an object onto the image sensor. Ideally, one point at the object is mapped to another point at the image plane. These two points are conjugate to each other, as shown in figure 5.1. So this is a point-to-point mapping for the imaging technology. The optical design task is to build up a perfect imaging system to form a tightening image point, and the location of the image point must be as accurate as possible. To do that, the designer has to conquer all the associated problems to make the image point perfectly. These problems include spatial aberrations of the optical components causing image blur, the chromatic aberration causing color distortion, and misalignment causing image degradation. In an imaging system, each point on the object can be regarded as a point source, so that the object contains a lot of point sources [3].

doi:10.1088/978-0-7503-2368-0ch5

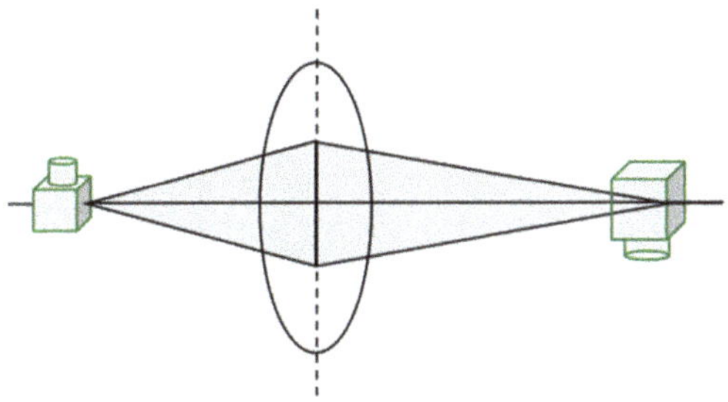

Figure 5.1. The lens is used to project an object to the image with point-to-point mapping.

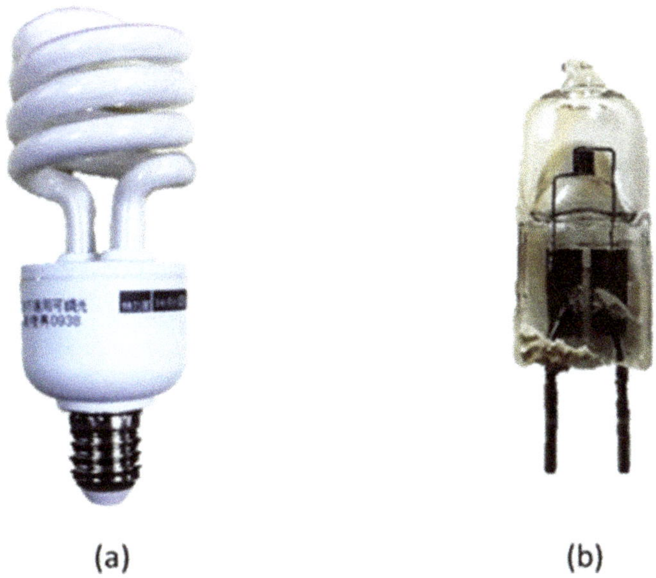

Figure 5.2. Two different light sources: (a) a compact fluorescent tube, and (b) a tungsten halogen.

However, most optical systems for lighting purposes are nonimaging rather than imaging [2]. The point-to-point mapping is usually not the issue in a nonimaging system, and this is the main difference between imaging and nonimaging systems. The design approach for a nonimaging system always relates to the light source, which emits the light and defines the light field through optical elements, including lens, reflectors and diffusers [4]. Generally, a light source will provide sufficient flux to the system, and the nonimaging system needs to manage the light field through designed optical elements. Therefore, the light source needs to be well modelled [5, 6]. There are various light sources, and different light sources could perform different light fields with different flux levels [7–9]. Figure 5.2 shows two different light sources, including a tungsten halogen and fluorescent tube, and anyone can distinguish one from another.

White LEDs are currently the most important light source through all the applications [10, 11]. In contrast to light sources through thermal radiation, an LED light source radiates light through radiation of luminescence with electron–hole

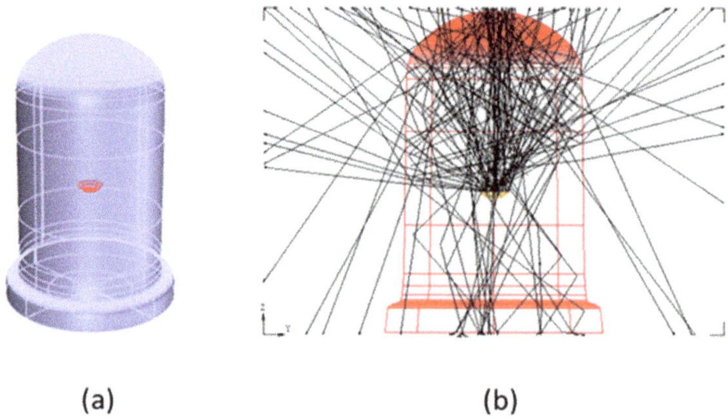

Figure 5.3. (a) The geometry of a bullet-shape LED, and (b) simulation of light emitting.

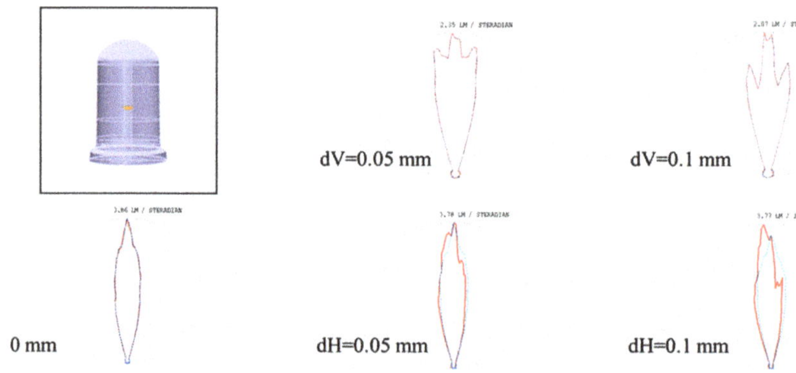

Figure 5.4. The simulated light patterns of a bullet-shape LED where lateral (dH) or vertical (dV) shift of the LED die is introduced.

pair combination, and a white LED needs an additional process of down conversion in the phosphor [12]. Even a white LED usually looks small, it is not a point source. Figure 5.3 shows an LED in a bullet shape, where an LED die is attached on the bottom surface of a reflective cup, and both are immersed in a bullet-shape epoxy volume [13]. Ray fan analysis shows that the light distribution is complicated. A precise light source model for such an LED must contain all the optical characteristics such as surface roughness and reflectivity, locations of those objects inside the epoxy volume and the refractive index and the curvature of the exit face of the epoxy. Dislocation of those objects will make the projection light pattern different, as shown in figures 5.4 and 5.5. These two figures from the simulation show an example to illustrate how the light field is affected by a light source when the components are not at the designed location. Thus one can imagine that a similar LED could project the light in a completely different way, just because of a slight deviation of a certain element from the design. The different light source behaviour will result in different illumination distribution on the target. This means that a precise light source model is important and necessary for lighting design.

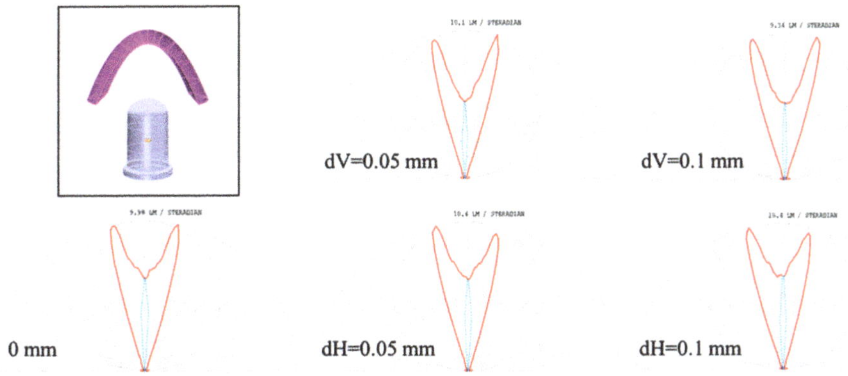

Figure 5.5. The simulated light patterns of a bullet-shape LED with a projection lens where lateral (dH) or vertical (dV) shift of the LED die is introduced.

5.2 The propagation fields

Lighting design is an approach to design optical elements to illuminate a target with a specific light source [1]. To figure out the effect and characteristic of the propagation distance to a light source, light field properties should be known [14, 15]. No matter how the light is radiated, the emitting light could propagate inside the light source and/or outside the light source. This condition is very special in an LED. The previous chapter has introduced total internal reflection that will stop emitting photons to escape from the LED crystal. Therefore, various approaches try to increase the light extraction efficiency from an LED [16–19]. For an LED, there could be more or fewer photons trapped in the LED crystal, and finally are absorbed so that the luminous efficacy decreases. However, when a photon hits the boundary of the LED crystal and is reflected through total internal reflection, the light field of the photon is not limited inside the LED volume. On the boundary of the LED crystal, there is an evanescent light field across the boundary. This means that outside the LED, the light field is there, but in an exponential-decay manner, as shown in figure 5.6. Therefore, even if the light is totally reflected by the boundary, the light field can penetrate the boundary at a very short distance and could be touchable if an object approaches the boundary at a distance short enough. Such a field is not in a propagation mode, so a user cannot utilize it. The range to confine the evanescent wave is called the near field [14]. Since the near field is always there, all the light is incident to this field. If a light passes through the field and becomes a propagation mode, it is useful for illumination. If light is stopped by the boundary, it could be an evanescent wave, and will not be propagated outside unless the surface or boundary condition is changed. There are various studies paying attention to how to change the boundary condition and transfer the evanescent wave to a propagation mode, so that the emission efficiency of a light source can be increased [20–24].

In most illumination scenarios, the illuminated target is usually located at a distance larger than ten times the largest lateral size of the light source, so the illuminated target is in the far field region of the light source [25]. In some research, the far field distance could start from five times the largest lateral size of the light

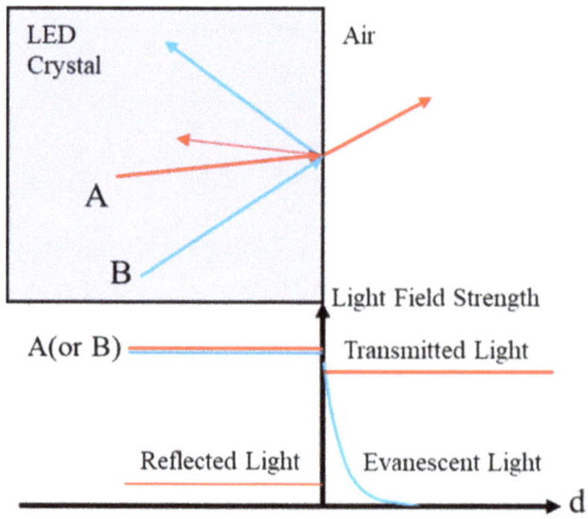

Figure 5.6. A schematic diagram to illustrate the light fields across the boundary. The blue light is reflected through total internal reflection and the red light is split into two parts of light, where one is reflected and the other is transmitted through Fresnel reflection. (d: the propagation distance from the boundary).

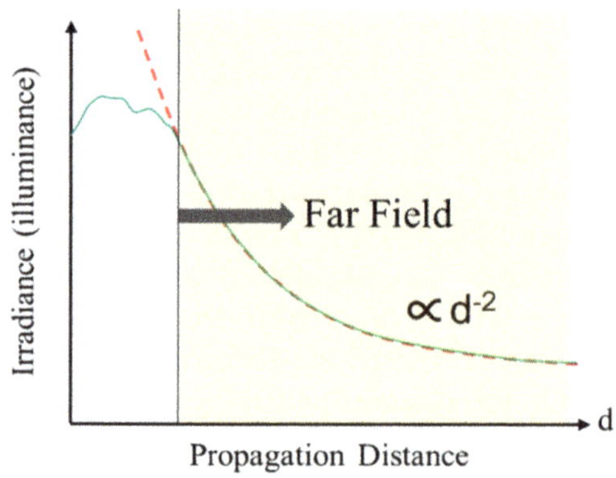

Figure 5.7. A way the define the far field region through inverse square of irradiance or illuminance via propagation distance. Green line: the measurement; red dash line: the line for inverse square function.

source [26–28]. The physics insight of the far field can be illustrated in different ways. One of them is that when an observer is in the far field region, the light source will look small, and the structure of the light source could not be resolved. Another important feature of the far field is that when the illumination is measured with intensity, i.e., flux per solid angle, the angular light pattern (intensity) will not change no matter how far the observer is. This means that the intensity distribution will be unchanged when the observer is in the far field. The third feature is that the measured illumination (irradiance) will inverse square proportional to the observation distance, as shown in figure 5.7, and this is a characteristic of a point source or a small light source [14, 25, 29, 30].

When the viewing distance is within 10 times the lateral size of the light source, the light field behaviour will be completely different from that in the far field [25].

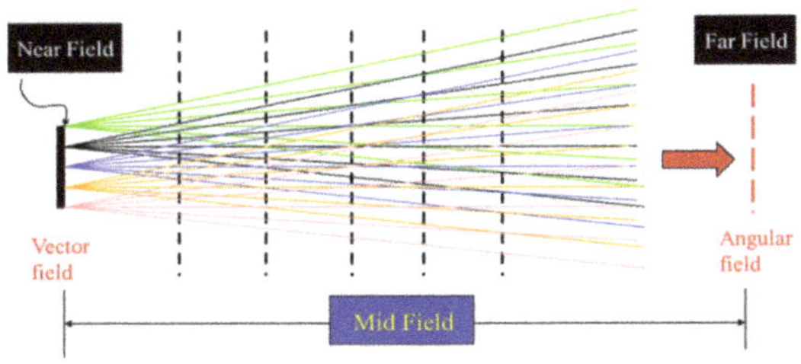

Figure 5.8. The definition of the light field in lighting. The near field contains non-propagation mode. The far field makes the intensity unchanged. The midfield is just between the near field and the far field. The light pattern changes quickly in the midfield until it reaches the far field.

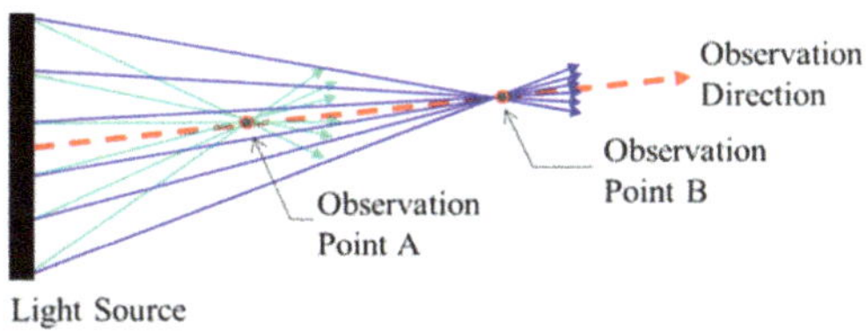

Figure 5.9. The feature of the midfield: the light field is not correlative between two points along the same observation direction.

As shown in figure 5.8, the whole field can be divided into three regions. The nearest field is called the near field, where the light field could contain an evanescent wave, which is not in the propagation mode. The furthest is the far field, where the light field is an angular field which will not change with the view distances. The other field is the region between the near field and the far field, and it is called midfield. In the midfield, the (angular) light pattern will change when the viewing distance changes. In this region, the light source does not look small, and the light source acts like an extended light source [15]. It means that for a certain viewing point, the received light field could not be similar with that in another viewing point along the same viewing direction, as shown in figure 5.9. The light field distribution is difficult to illustrate using a simple formula.

The distance of the far field is not as far as one might imagine [25, 29]. For example, for a white LED with an effective emission surface of 2 mm × 2 mm, the far field could start from 1 cm or 2 cm from the light source. It is quite a short distance. Thus the midfield means no effect on a white LED when used in illumination. However, the condition is changed when the light source is changed from a white LED to a light source module, where a lens or a reflector redirects the light field. Then the lateral size of the light source will be changed to the lateral dimensions of the lens or reflector. If the lateral dimensions are 25 mm × 25 mm, the midfield could be extended to 125 mm or 250 mm. Then the operation distance of a user could fall into the midfield region. If several modules are contained in a luminaire, the effective lateral dimensions should be the whole width of the luminaire, as shown in figure 5.10. Note that figure 5.10 illustrates a general condition for the light module with a wide large divergent angle, such as a

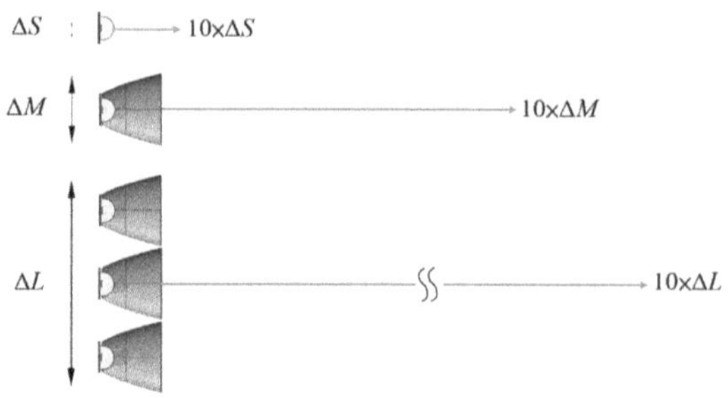

Figure 5.10. The midfield range is within ten times of the lateral dimension of the light source, where ΔS, ΔM and ΔL are the lateral dimensions of the LED, LED module and the luminaire, respectively.

Lambertian light source. When the divergent angle of the effective light source is not as wide as that of a Lambertian light source or the separation distance between each two modules is too long, the midfield could be longer than that shown in figure 5.10.

A good example to illustrate the characteristic of the midfield is a reading light, where the working distance is within 30 cm to 100 cm, but the luminaire width could be as long as 20 cm. If a user's eyes are located at the far field of the luminaire, the light pattern will not change a lot when the user's eyes move in a limited range along the vertical direction. However, it is possible that a user's eyes fall into the midfield region. Then it is necessary to figure out the characteristics clearly. To illustrate the divergent angle of a light source, it is convenient to describe the intensity as function of viewing angle

$$I(\theta) = I_0 \cos^m(\theta), \tag{5.1}$$

where I_0 is the luminous intensity at the normal direction, m is the cosine power factor, and θ is the viewing angle. Equation (5.1) describes a Lambertian light source when $m = 1$, a point source when $m = 0$, and a collimated light m approaches infinity [31, 32], as shown in figure 5.11. Note that equation (5.1) is for the far-field pattern, and the effective light source (or the LED) is a point source or said small light source. To do it in a more practical way, four different reflectors with a certain white LED shown in figure 5.12 are used to perform four different light patterns, where the FWHM angles are 30°, 45°, 60°, and 80°, respectively [32]. Figure 5.13 shows a luminaire, where 4x3 LED modules are contained. The simulation of the central illuminance and the illumination pattern at different viewing distances are shown in figure 5.14. The simulation shows that under the same geometry, the divergent angle of the LED module plays a role in determining the distance of the far field. The larger the divergent angle is, the shorter the starting point of the far field will be. When the central illuminance approaches the curve by inverse square law, the light pattern is more uniform and it means that the viewing distance is in the far field. The simulation shown in figures 5.13 and 5.14 applies the models shown in figure 5.12, which is precise but complicated. When the light source module is replaced by a point source with the far-field intensity distribution as described in equation (5.1), it is easier to calculate the central illuminance via the viewing distance. A comparison between the simulation with the point source and LED module is shown in

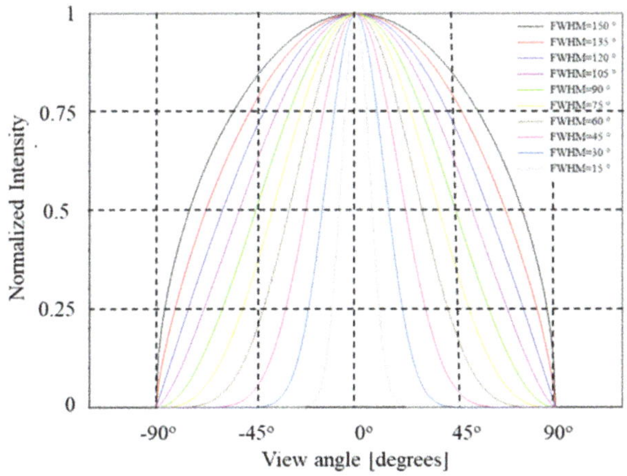

(a)

FWHM	m
150°	0.51
135°	0.72
120°	1.00
105°	1.40
90°	2.00
75°	2.99
60°	4.82
45°	8.75
30°	19.99
15°	80.67

(b)

Figure 5.11. (a) The intensity versus viewing angle of a light source, and (b) the FWHM angle and the corresponding power m in equation (5.1).

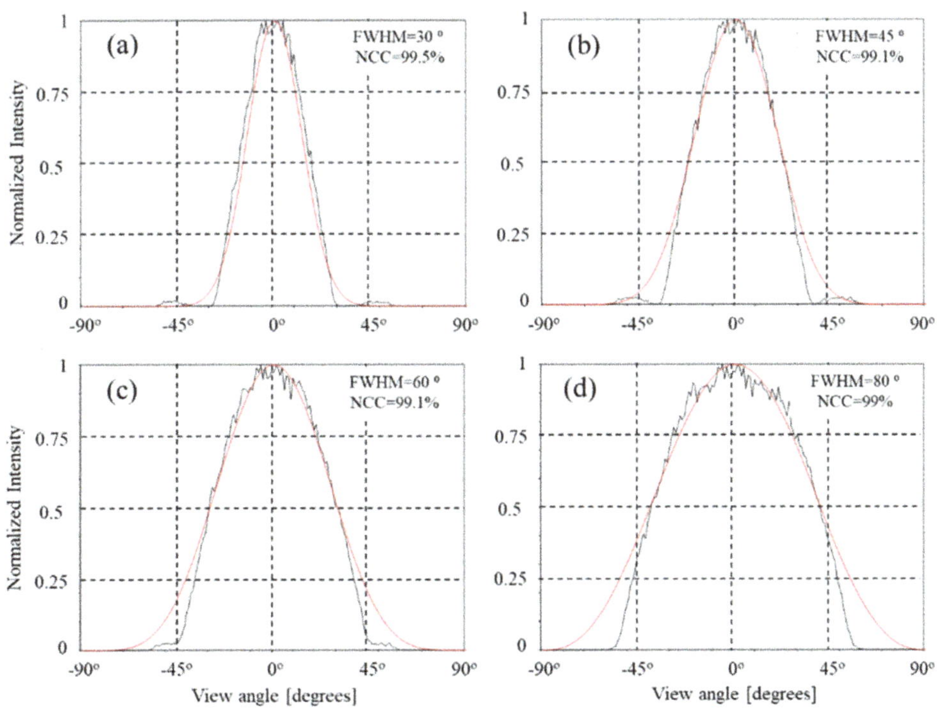

Figure 5.12. Four different reflectors with a white LED are used to perform four different light patterns, where the FWHM angles are (a) 30°, (b) 45°, (c) 60°, and (d) 80°.

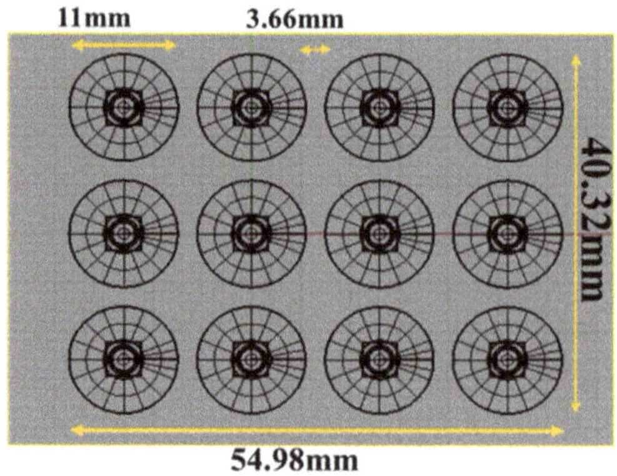

Figure 5.13. A luminaire contains 12 LED modules.

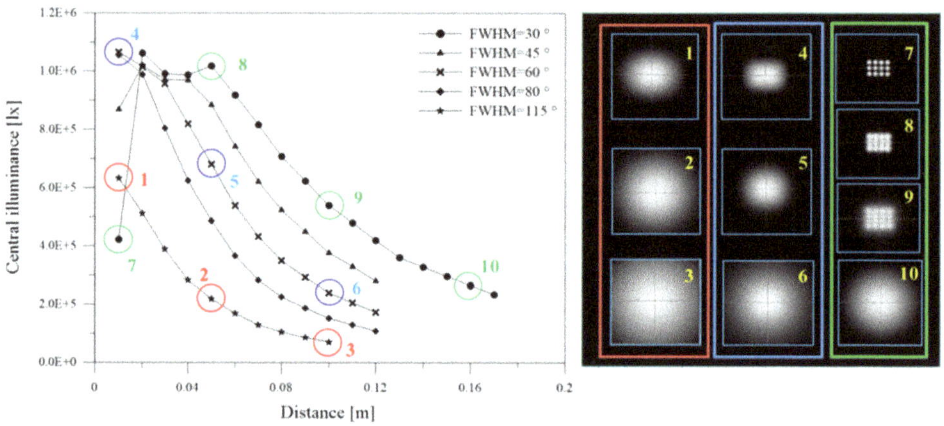

Figure 5.14. The central illuminance as a function of viewing distance.

figure 5.15, where R_{FF} is the starting distance of the far field, and D is the largest lateral dimension of the luminaire. In contrast to a complicated model for the LED module, a simple far-field model of the LED module is useful to predict the light field pattern of the luminaire near the far field [32].

Figure 5.16 shows an interesting example, where all the scales are enlarged to clearly see the difference between the midfield and the far field. The luminaire is composed of nine sets of the secondary LED luminaire and each secondary LED luminaire contains 16 pieces of LED module. Thus, a total of 144 LED modules construct the large-size LED luminaire. From the close view of the light field, the individual beams are visible, and all the individual beams merge to one at a certain distance, which is the starting distance of the far field. When an observer looks at the upward beam by the luminaire at a far distance, the upward beam is a single beam because what the observer can see is the far-field beam pattern. This is one of the important features in the far field, and the individual beams stand for the midfield light pattern [33].

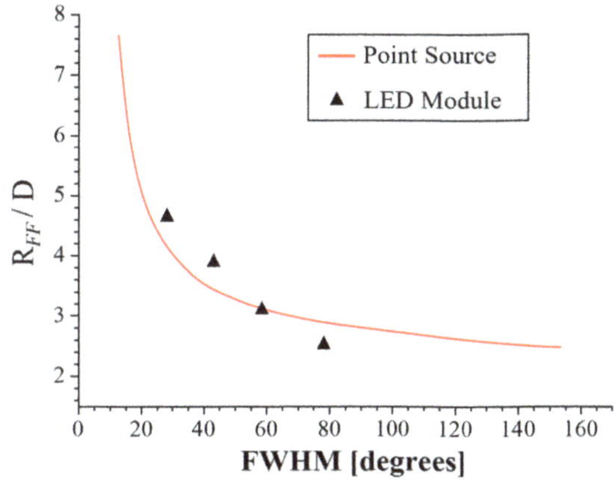

Figure 5.15. Simulation of the starting distance of the far field versus the FWHM angle by two approaches, one is with equation (5.1) for point sources and the other is with LED modules.

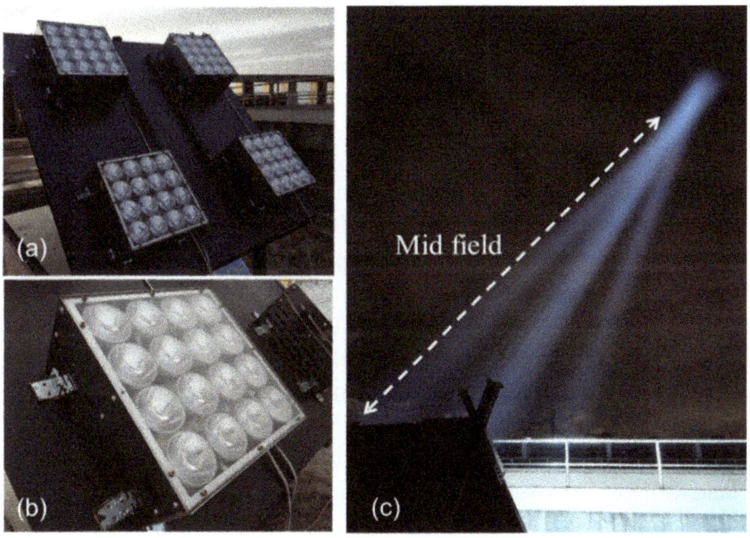

Figure 5.16. (a) The luminaire contains four secondary LED luminaires; (b) each LED luminaire contains 16 modules; (c) the light beams do not merge until the propagation distance reaches far field.

In the lighting design and illumination scenarios, the midfield is a region playing important roles, especially in light source modelling [15, 25]. The simplest model is described in equation (5.1), where the light field is located at the far field region, so it is applicable to describe the light field or illumination in the far field. An alternative way is to build a simple light source model in a commercial simulation software by tracing the light field from the far field to the light source which can still be treated as a point source [30, 31]. However, such a light source model cannot perform the precise light distribution in the mid field [15, 34, 35]. In contrast, if the light source model is precise enough, it should provide the accurate light distribution in all fields that the emission light is in the propagation mode [31].

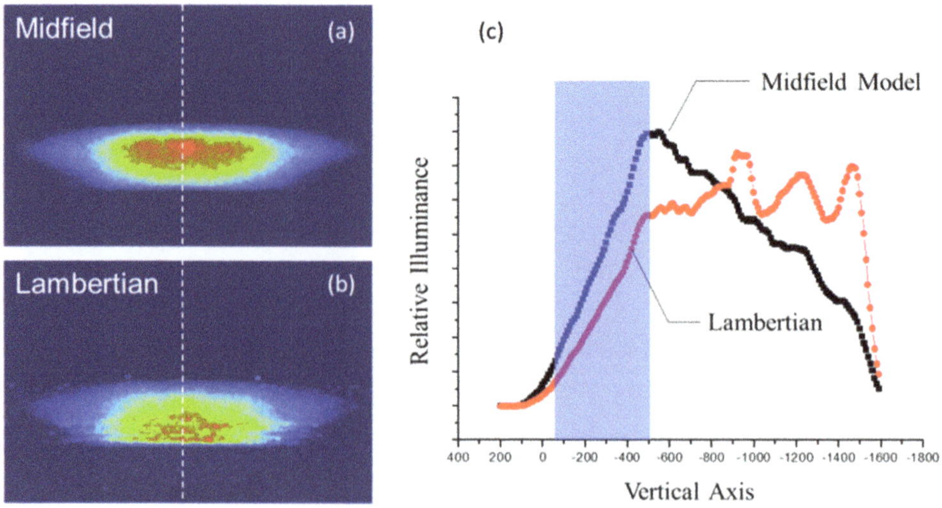

Figure 5.17. A comparison between a midfield model and a simple model with a Lambertian light source. (a) The simulation with the midfield model, (b) with the Lambertian light source, and (c) the simulated illuminance versus vertical axis on the target plane 10 m away from the light source. The blue region is for the high-contrast cut-off line.

So why is the midfield model important? We know that the midfield covers the distance from hundreds of micrometers to ten times the lateral size of the light source [25]. In LED optics, an important task is to design the optics to project the light onto the target area with predicted quality. It is easy to understand that all the possible optical components are located in this region. If the midfield model is incorrect, there must be a certain error of the light field distribution across the optical components. Figure 5.17 shows an example. The correct model is precise and can accurately simulate the light field distribution in the midfield. The incorrect model is to use a Lambertian light source to replace the real light source. Even though the area of the Lambertian light source is the same as that of the real light source, the light field distribution across the midfield is different from that of the real light source. Through an optical element, the light pattern at 10 m will be different. Such a difference could make an optical design fail to meet the requirement. Consequently, the projected light pattern in the design is not trustable.

5.3 LED light source modelling

To have a precise light source model is essential in lighting design [5, 6]. Generally, there are two modelling approaches. One is to catch the image of the light source from multiple views. Each frame corresponds to a viewing angle, and a specific propagation direction of the light from vision at the far field. Then image fusion technology is applied to construct a precise light source model, which is useful to predict the light field with propagation modes in the midfield and far field as well. Such a technique is an effective way to generate a precise light source model, but the disadvantage is that the model loses freedom to modify the structure of a light source

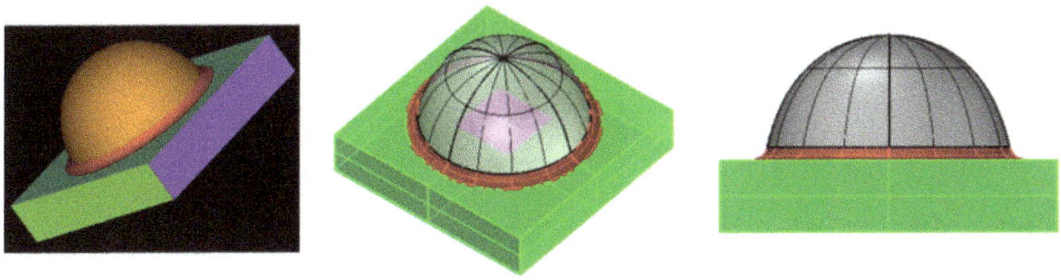

Figure 5.18. Three viewing angles of the LED geometrical model.

when some components of the light source are adjustable. The condition becomes worse when the technique is applied to a light source designer who needs to modify the detailed structure of the light source, such as in the packaging development process. In addition, such an approach needs to equip an expensive instrument.

The other approach is to build up a geometrical model according to the real light source [15]. The concept of the approach is different from the imaging fusion technique, but is more flexible in developing a light source. Figure 5.18 shows an example of a white LED, where the location of the LED die and all the other components are precisely built. Besides, all surfaces inside the light source are well defined with appropriate surface property, such as scattering property, reflectivity and color.

The emission characteristic of the LED die is one of the key issues [15]. Essentially, a designer needs to trace the light from the active layer, because all the rays are emitted from the active layer [36]. But it is time consuming if all the propagation-mode rays are traced from the active layer. Instead, the designer could simulate the emission characteristic of all the effective exit surfaces of the LED die. In a monocolor LED, there are five faces needing to be taken care of. Keep in mind that only a part of the rays can escape from the LED die because there is total internal reflection to trap lots of light emitted from the active layer. Therefore, the simulation could use the effective exit surface to record the light emission characteristic of the light field, and the five exit surfaces become new light sources. In the final model, the rays are emitted by the effective surfaces rather than from the active layer [15]. If the modelled LED is a white LED, the exit surfaces could be covered with phosphor layers. The optical property of the phosphor layer has been introduced in the previous chapter. Generally, a phosphor layer can be regarded as a Lambertian surface. Thus the LED die with phosphor coating is easy to model [37].

Applying Monte Carlo ray tracing with a commercial simulation program is the best way to construct the model with precise light field distribution in the midfield and far field [5, 6, 15, 38, 39]. To verify the accuracy of the light source model, one-dimensional angular radiation pattern (or intensity) measurement is necessary [31], as shown in figure 5.19(a). The real light source is mounted at the rotational center of a rotator. A photo detector with a tiny aperture is on the rotation plane to detect the angular radiation pattern when the rotator is rotated from −90° to 90°. To make sure that the measurement distance is located at the midfield, the photo detector must face the emitter as close as possible. In the corresponding simulation, the

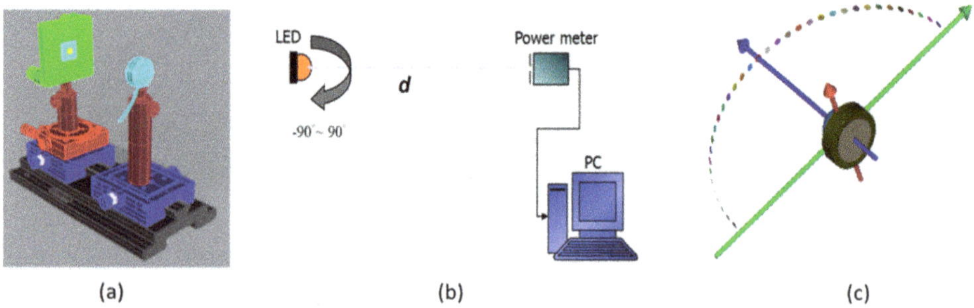

Figure 5.19. The geometry of angular radiation pattern measurement. (a) A 3D layout, (b) a schematic diagram, and (c) the corresponding simulation layout.

detectors are located around the emitter at the same distance to catch the angular light pattern, as shown in figure 5.19(b) [40]. Three angular light patterns at three measurement distances are suggested to compare with the simulation at the same distances. Then a comparison between the measured and simulated angular patterns can be evaluated by normalized cross correlation (NCC), which is written

$$\mathrm{NCC} = \frac{\sum_x \sum_y (A_{xy} - \bar{A})(B_{xy} - \bar{B})}{\sqrt{\sum_x \sum_y (A_{xy} - \bar{A})^2 \sum_x \sum_y (B_{xy} - \bar{B})^2}}, \tag{5.2}$$

where A_{xy} and B_{xy} are the intensity or irradiance of the simulation (A) and experimental values (B); $\bar{A}$ is the mean value of A, and $\bar{B}$ is the mean value of B [15, 41]. The whole procedure is shown in figure 5.20. The modelling procedure starts from determining the LED parameters, including location, dimension, transmission efficiency and reflectivity. Then emit rays from the active layer with Monte Carlo ray tracing. The ray vector or the light field is recorded on each effective exit surface of the LED die. Then sufficient ray numbers in the simulation must be done [15]. Figure 5.21 shows the effect by the ray number in the simulation. Generally, more than ten million rays are necessary to precisely simulate a light pattern, but only ten thousand rays are enough to calculate the efficiency. The ray number is not sufficient when the simulated light pattern looks similar to a speckle pattern. Then the light field distribution in the midfield and the far field can be obtained. Since the light pattern in the midfield changes from one distance to another, three simulated patterns at different distances are enough to judge if the model is accurate. However, the angular light pattern in the far field is the same. It is possible to have the same far-field light pattern but with a different midfield light pattern. Therefore, the comparison of the light pattern in the far field is incorrect. In the midfield model, the simulated midfield pattern must be compared with the measurement of the real sample at the same distances [15, 40]. Through the calculation of NCC shown in equation (5.2), similarity between the simulation and the measurement can be obtained. In the reports [42], a precise midfield model is completed if the NCCs reach 99.5% or above at two different measurement distances in the midfield, as shown in figure 5.22.

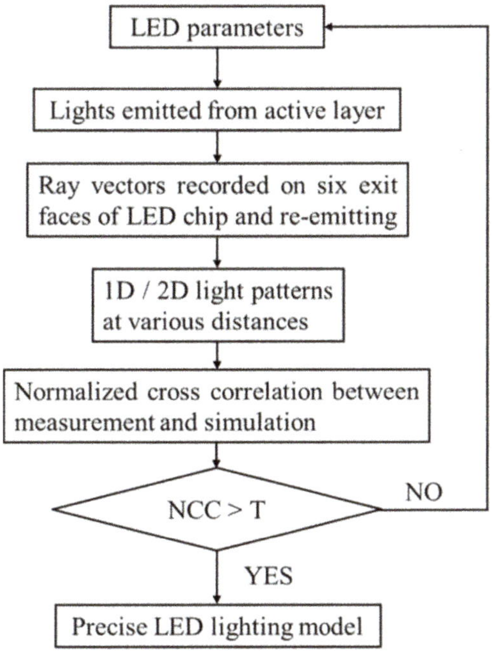

Figure 5.20. The procedure of midfield model.

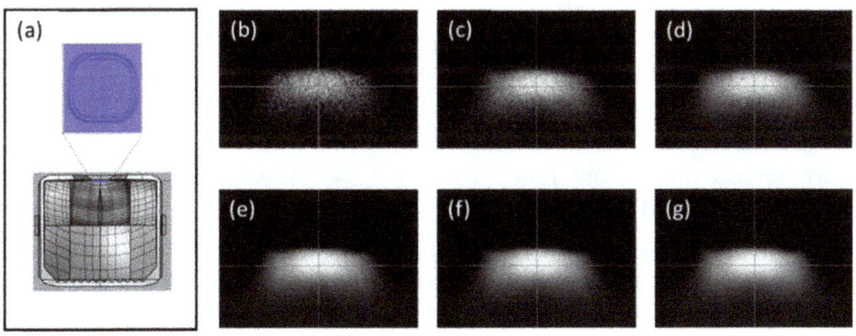

Figure 5.21. (a) A reflector with a white LED used in the simulation. The simulated light pattern has a ray number of (b) 100 000, (c) 500 000, (d) 1 000 000, (e) 5 000 000, (f) 10 000 000, and (g) 20 000 000.

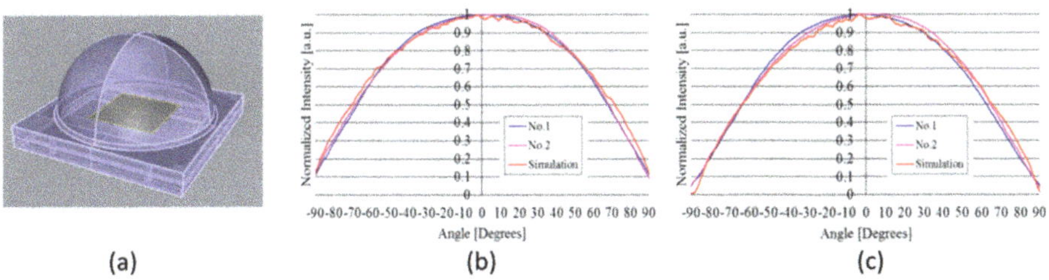

Figure 5.22. (a) The geometry of the LED, where the lateral dimensions are 5 mm × 5 mm, and the height is 2.95 mm; the two measurements and the simulated light patterns at (b) 15 mm, and (c) 30 mm. The NCC is 99.5% at 15 mm, and 99.6% at 30 mm.

If the midfield model is not necessary in optical design or illumination analysis, the modelling procedure is much easier. The first approach is to measure the light radiation patterning the far field, and then trace back to the point light source. Most Monte Carlo ray tracing programs have such a function. A useful illumination software named DIALux is a very powerful simulation program based on a far field model [43]. Another powerful model without a ray tracing program is to build up a simple function such as equation (5.1) or a complex function with special polynomials or functions to describe the far-field pattern. The far-field radiation pattern should be in three dimensions. A flexible hybrid function proposed by Moreno and Sun is a powerful way to predict the far-field pattern and is helpful to an optical designer to figure out a complicated illumination system in the far field [15].

5.4 Case studies of light source modeling

In this section, we will introduce the ways to model the light source in different types. The first is a single-color LED, the second is phosphor-converted white LED (so-called pcW-LED), and the third is a pcW-LED with multiple LED dies.

A single-color LED is without phosphor coating, so that the color of the LED is the same as that emitted by the LED die. The light source model relates to the dimensions of the LED die and the packaging structure as well. For example, the structure of a bullet-shape LED is different from an LED with surface mounted technology (SMT). To model the light source, we need to know the emission characteristics of the five effective emission surfaces of the LED die, as indicated in the modelling procedure. Figure 5.23 shows the geometry of a bullet-shape LED, where the reflectivity of the reflective cup, and the refractive index as well as the absorption of the epoxy lens have been known. Once the five emission surfaces are figured out, more than ten millions of rays are sufficient to have a stable light distribution across the whole field. Figure 5.24 shows comparisons between the simulated and measured patterns at three different locations in the midfield. The NCC is 97.7% at 1.5 cm, 98.2% at 3 cm, and 98.5% at 5 cm [13].

In an SMT LED, the five emission surfaces are the most important factors in the modelling procedure. Figure 5.25 shows figures to check the emission characteristic of the top surface and side surfaces of a blue die in an SMT LED. If the thickness of an LED is not neglectable, the emission characteristics of the four sides are

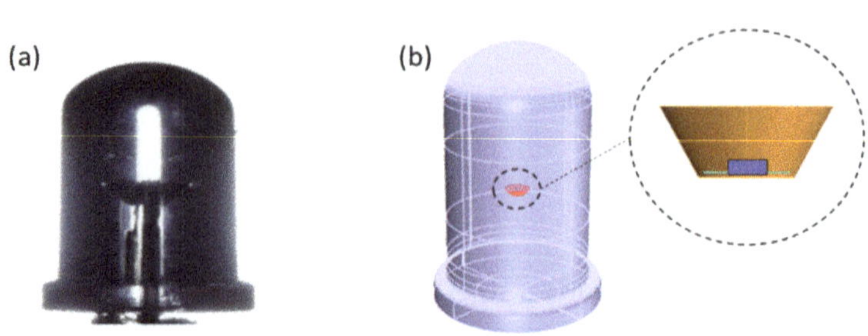

Figure 5.23. (a) A photo of the bullet-shape LED (reprinted with permission from [47]), and (b) the corresponding geometry.

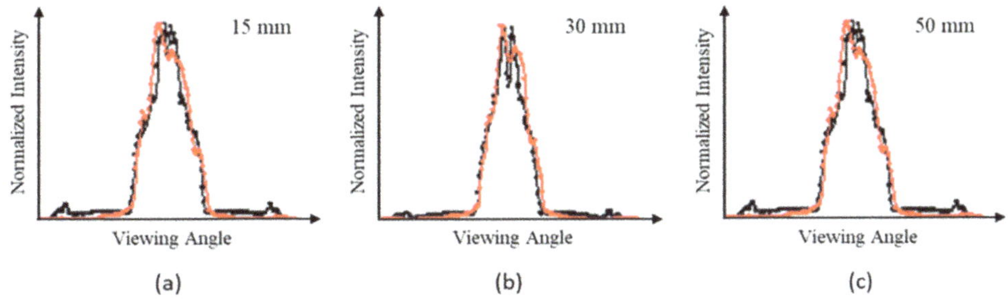

Figure 5.24. The comparison between the simulated and measured light patterns at the distance of (a) 15 mm, (b) 30 mm and (c) 50 mm. Reprinted with permission from [47].

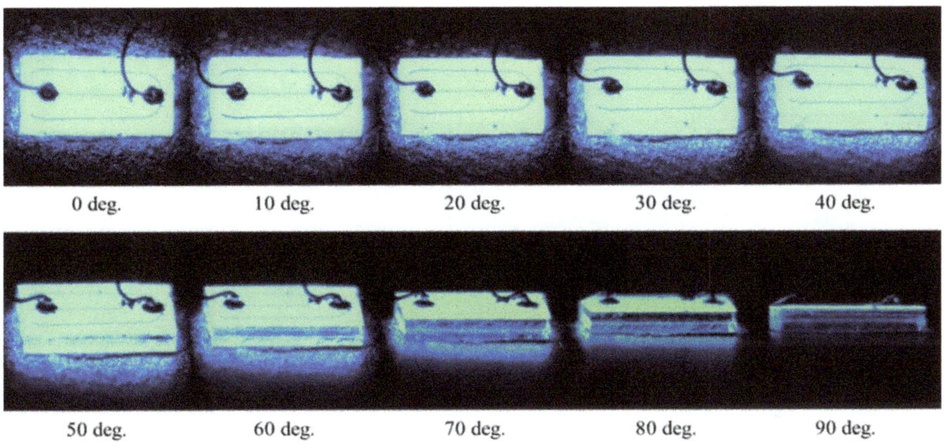

Figure 5.25. The photos at different view angles to an LED die with large thickness. Reprinted with permission from [48].

important. Besides, the reflectivity of the bottom surface in the SMD (surface mounted device) could play an important role because the reflected light from the bottom surface cannot be neglected. Figure 5.26 shows the simulation result by the constructed model following the modelling procedure shown in figure 5.20. A series of photos are taken to figure out the emission characteristics of the four sides of the LED die. The emission characteristic in the four sides is the key factor to the optical model. Generally, the emission from each side is not like a Lambertian surface, so that more experimental measurement such as figure 5.25 is necessary to figure out the emission property [44].

To model a pcW-LED is a little bit different from a single-color LED because the phosphor coating could decide the surface emission characteristic of the light source. Generally, phosphor coating is a way to spread a uniform layer of phosphor on the LED die. The phosphor has two roles in a white LED. One is to re-emit light at a

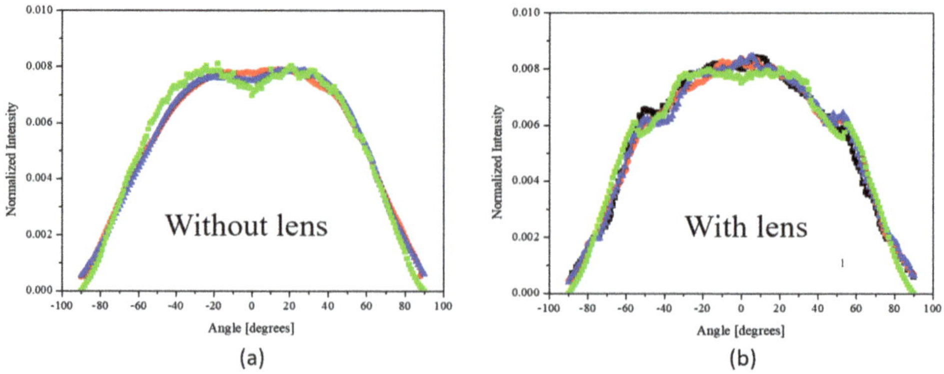

Figure 5.26. The simulated and measured radiation pattern at a distance 20 cm. (a) The LED is without encapsulation, and (b) the LED is with a lens encapsulation, where the green lines are from simulation.

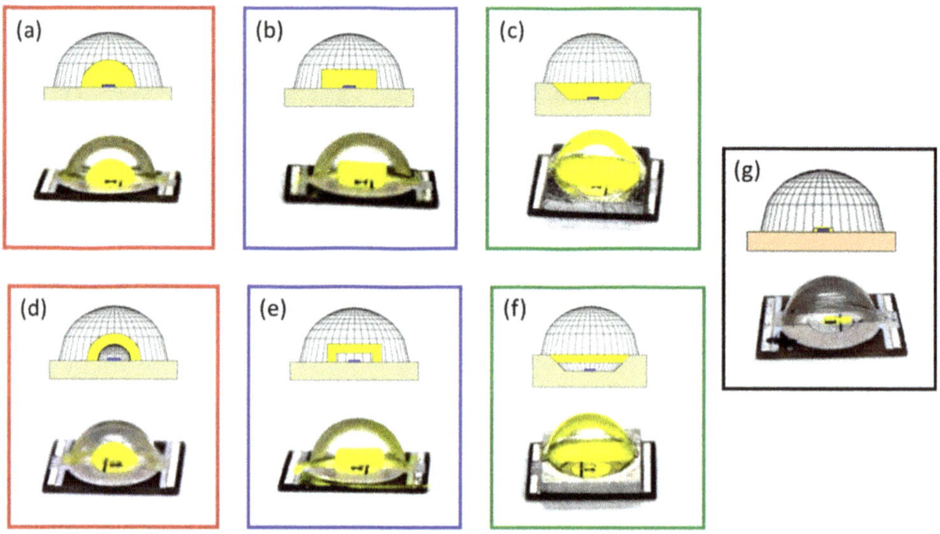

Figure 5.27. Different types of packaging geometry. (a) Phosphor hemisphere, (b) volume cube, (c) dispensing into a cup, (d) a remote phosphor type of (a), (e) a remote phosphor type of (b), (f) a remote phosphor type of (c), and (g) conformal coating. Reprinted with permission from [49].

longer wavelength through down conversion of the phosphor. The other is to scatter the blue light which is not absorbed by the phosphor. As a result, the white light can be regarded as a Lambertian emitter in most cases. Thus, a pcW-LED is easier to model than a single-color LED. Figure 5.27 shows several typical phosphor coating methods, where the conformal coating is the most popular in industry products for its advantages in cost, optical property and color property [45].

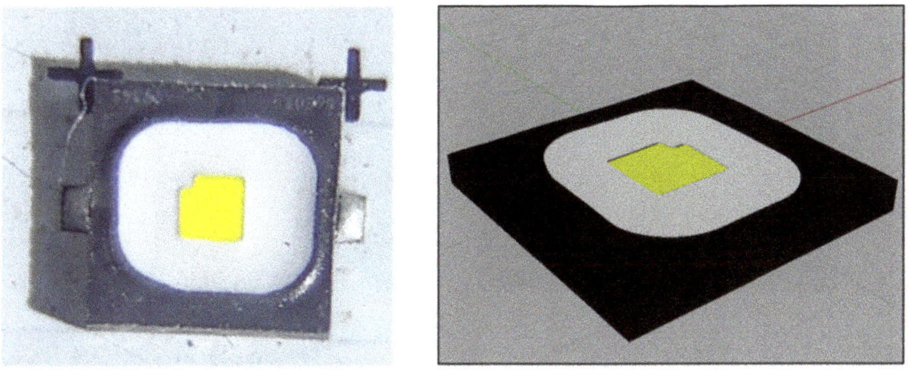

(a) (b)

Figure 5.28. (a) A white LED with conformal coating, and (b) its geometry model (reprinted with permission from [50]).

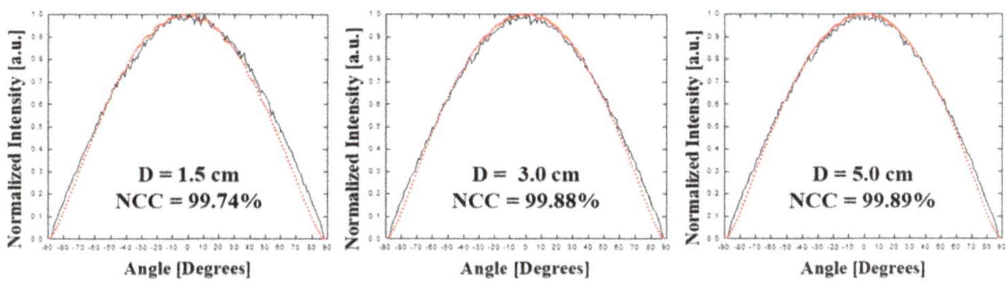

(a) (b) (c)

Figure 5.29. The comparison between the simulated light pattern and the measurement at different distances, (a) 1.5 cm, (b) 3 cm, and (c) 5 cm. Reprinted with permission from [50].

Figure 5.28 shows a typical SMT pcW-LED and its geometrical model. The phosphor coating in the pcW-LED is through conformal coating, where only the LED die is coated with phosphor layer. In most cases, the conformal coating will make the emission from a Lambertian surface. The simulation based on this assumption is shown in figure 5.29, where the NCC is always larger than 99% in the mid field and the far field [33].

Even if the conformal coating is applied, the model could be modified when there are multiple dies in a single package volume. Figure 5.30(a) shows a multi-die pcW-LED, where two LED dies are in a package volume with phosphor coating [46]. The model becomes more complicated because the scattering light will laterally spread in the phosphor layer outside the LED die. Thus the brightness across the phosphor layer is not homogeneous, as shown in figure 5.30(b). The way to solve the problem is to catch the surface image with a mono-color camera to analyze the brightness level across the phosphor layer, and put a weighting factor to indicate the brightness level of a specific region of the phosphor. The result will make the light source model more precise.

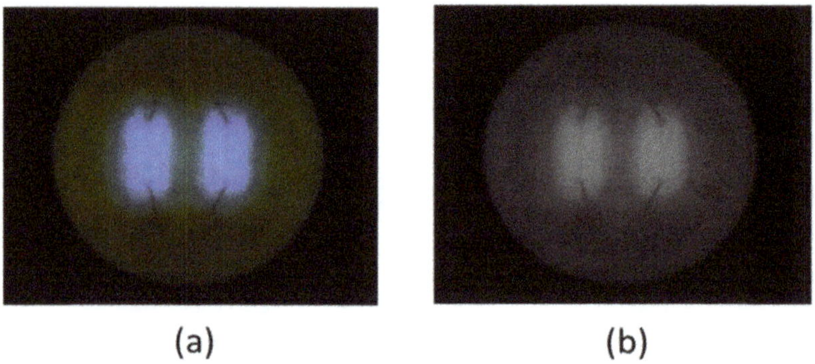

Figure 5.30. (a) A photo of a white LED with two dies, and (b) the photo in black-and-white. Reprinted with permission from [51].

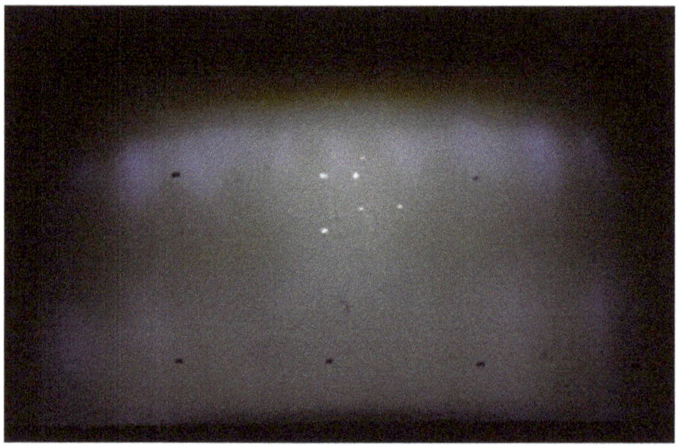

Figure 5.31. A photo of the illumination pattern at a distance 10 m away from the headlamp, where yellowish and bluish pattern can be seen. Reprinted with permission from [51].

Sometimes, for the application to a high-precision luminaire such as a vehicle headlamp, the modelling procedure is further complicated to ensure the validity in color as well. Figure 5.31 shows an illumination pattern on a wall at a distance 10 m from a headlamp with use of the light source shown in figure 5.30. The radiation pattern is yellowish in some areas and bluish in others. Such a phenomenon is caused by color difference across the phosphor layer in the pcW-LED. Since the pcW-LED mainly contains blue light and yellow light, taking a photo with a blue or yellow color filter is useful to know the brightness level of each color, as shown in figure 5.32. Thus a final model should consist of light source models with precise weighting factor to describe the color deviation of the projected light pattern. Figure 5.33 shows the simulation by the two models, and the difference between the blue pattern and the yellow pattern is obvious [46].

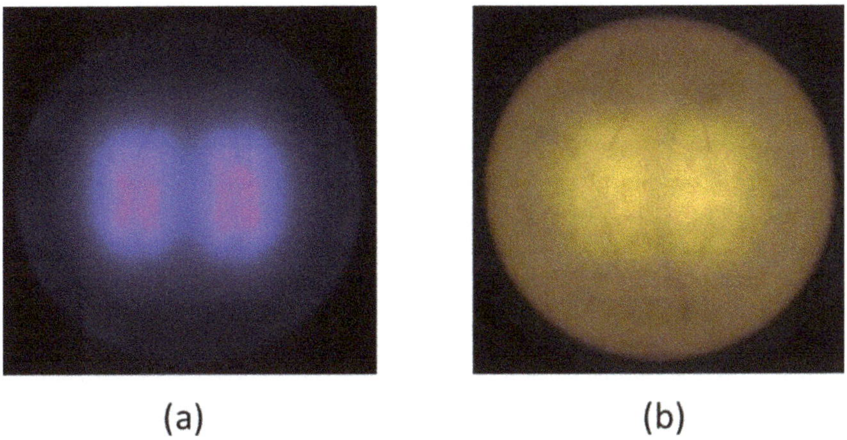

Figure 5.32. The photos of the white LED shown in figure 5.31 with (a) a blue filter, and (b) a yellow filter. Reprinted with permission from [51].

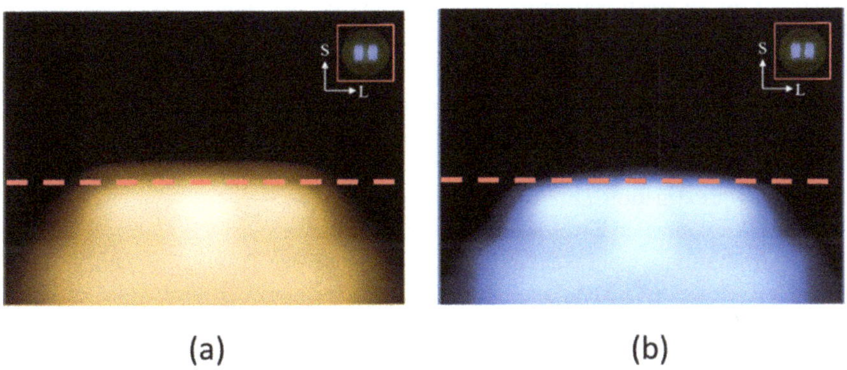

Figure 5.33. The photos of the illumination pattern at distance 10 m, with (a) the yellow model, and (b) the blue model. Reprinted with permission from [51].

References

[1] Yang H, Bergmans J W M, Schenk T C W, Linnartz J P M G and Rietman R 2008 An analytical model for the illuminance distribution of a power LED *Opt. Express* **16** 21641–6
[2] Benitez P and Miñano J C 2007 The future of illumination design *Opt. Photon. News* **18** 20–5
[3] Mahajan V N 1998 *Optical Imaging and Aberrations, Part I: Ray Geometrical Optics* (Bellingham, WA: SPIE Press)
[4] Winston R, Miñano J C and Benítez P G 2005 *Nonimaging Optics* (Amsterdam: Elsevier)
[5] Kaminski M S, Garcia K J, Stevenson M A, Frate M and Koshel R J 2002 Advanced topics in source modeling *Proc. SPIE* **4775** 46–57
[6] Zerfhau-Dreihöfer H, Haack U, Weber T and Wendt D 2002 Light source modeling for automotive lighting devices *Proc. SPIE* **4775** 58–66
[7] Schubert E F and Kim J K 2005 Solid-state light sources getting smart *Science* **308** 1274–8
[8] Narukawa Y 2004 White-light LEDs *Opt. Photon. News* **15** 24–9

[9] Zukauskas A, Schur M S and Gaska R 2002 *Introduction to Solid State Lighting* (New York: Wiley-Interscience)

[10] Nakamura S, Pearton S and Fasol G 2000 *The Blue Laser Diode: The Complete Story* (Berlin: Springer)

[11] Krames K R, Shchekin O B, Mueller-Mach R, Gerd G O, Ling Z, Harbers G and Craford M G 2007 Status and future of high-power light-emitting diodes for solid-state lighting *IEEE J. Displ. Technol* **3** 160–75

[12] Nakamura S, Mukai T and Senoh M 1994 Candela-class high-brightness InGaN/AlGaN doubleheterostructure blue-light-emitting diodes *Appl. Phys. Lett.* **64** 1687

[13] Lee T L, Chen Y C, Tsai M S and Sun C C 2017 Optical modeling of bullet-shaped LED for use in self-luminous traffic signs *Proc. SPIE* **10375** 103750

[14] Goodman J W 1996 *Introduction to Fourier Optics* 2nd edn (New York: McGraw-Hill)

[15] Sun C C, Lee T X, Ma S H, Lee Y L and Huang S M 2006 Precise optical modeling for LED lighting verified by cross correlation in the midfield region *Opt. Lett.* **31** 2193–5

[16] David A, Benisty H and Weisbuch C 2007 Optimization of light-diffracting photonic-crystals for high extraction efficiency LEDs *J. Display Technol.* **3** 133–48

[17] McGroddy K, David A, Matioli E, Iza M, Nakamura S, DenBaars S, Speck J S, Weisbuch C and Hu E L 2008 Directional emission control and increased light extraction in GaN photonics crystal light emitting diodes *Appl. Phys. Lett.* **93** 103502

[18] Bergenek K, Wiesmann C, Zull H, Wirth R, Sundgren P, Linder N, Streubel K and Krauss T F 2008 Directional light extraction from thin-film resonant cavity light-emitting diodes with a photonic crystal *Appl. Phys. Lett.* **93** 231109

[19] Wiesmann C, Bergenek K, Linder N and Schwarz U T 2009 Photonic crystal LEDs—designing light extraction *Laser Photon. Rev.* **3** 262–86

[20] Schnitzer I, Yablonovitch E, Carneau C, Gmitter T J and Scherer A 1993 30% External quantum efficiency from surface textured, thin-film lightemitting diodes *Appl. Phys. Lett.* **63** 2174–6

[21] Windisch R, Rooman C, Meinlschmidt S, Kiesel P, Zipperer D, Döhler G H, Dutta B, Kuijk M, Borghs G and Heremans P 2001 Impact of texture-enhanced transmission on high-efficiency surface-textured light-emitting diodes *Appl. Phys. Lett.* **79** 2315–7

[22] Huh C, Lee K S, Kang E J and Park S J 2003 Improved light-output and electrical performance of InGaNbased light-emitting diode by microroughening of the p-GaN surface *J. Appl. Phys.* **93** 9383–5

[23] Fujii T, Gao Y, Sharma R, Hu E L, Danbaars S P and Nakamura S 2004 Increase in the extraction efficiency of GaN-base light emitting diodes via surface roughening *Appl. Phys. Lett.* **84** 855–7

[24] Windisch R, Rooman C, Dutta B, Knobloch A, Borghs G, Döhler G H and Heremans P 2002 Light-extraction mechanisms in high-efficiency surface-textured light-emitting diodes *J. Select. Topics Quantum Electron.* **8** 248–55

[25] Sun C C, Chien W T, Moreno I, Hsieh C C and Lo Y C 2009 Analysis of the far-field region of LEDs *Opt. Express* **17** 13918–27

[26] IESNA 1985 *Photometric Testing of Indoor Fluorescent Luminaires*, IES LM-41–1985 Illuminating Engineering Society of North America

[27] IEEE 2000 *The Authoritative Dictionary of IEEE Standards Terms* 7th edn (Piscataway, NJ: IEEE)

[28] Ryer A 1998 *Light Measurement Handbook* (Newbury Port, MA: International Light) (http://intl-light.com/handbook/)

[29] Moreno I, Sun C C and Ivanov R 2009 Far-field condition for light-emitting diode arrays *Appl. Opt.* **48** 1190–7
[30] Moreno I and Sun C C 2008 LED array: where does far-field begin? *Proc. SPIE* **7058** 70580
[31] Moreno I and Sun C C 2008 Modeling the radiation pattern of LEDs *Opt. Express* **16** 1808–19
[32] Sun C C, Lin Y S, Yang T H, Lin S K, Lee X H, Wu C S and Yu Y W 2020 Illuminance and starting distance of the far field of LED-array luminaire operated at short working distance *Crystals* **10** 360
[33] Wu C S, Chen K Y, Lee X H, Lin S K, Sun C C, Cai J Y, Yang T H and Yu Y W 2019 Design of an LED spot light system with a projection distance of 10 km *Crystals* **9** 524
[34] Moreno I 2006 Spatial distribution of LED radiation, *Proc. SPIE* **6342** 634216
[35] Manninen P, Hovila J, Kärhä P and Ikonen E 2007 Method for analysing luminous intensity of light-emitting diodes *Meas. Sci. Technol.* **18** 223–9
[36] Zukauskas A, Shur M S and Caska R 2002 *Introduction to Solid-State Lighting* (New York: Wiley) ch 5
[37] Sun C C, Chen C Y, He H Y, Chen C C, Chien W T, Lee T X and Yang T H 2008 Precise optical modeling for silicate-based white LEDs *Opt. Express* **16** 20060–6
[38] Jongewaard M 2002 Guide to selecting the appropriate type of light source model *Proc. SPIE* **4775** 86–98
[39] Cassarly W J 2002 LED modelling: pros and cons of common methods *Photon. Tech Briefs IIa-2a*, special supplement to NASA Tech Briefs
[40] Sun C C, Lee T X, Ma S H, Lee Y L and Huang S M 2006 Optical modeling for LED in mid-field region,' in *Int. Optical Design Conf. J. Opt. Soc. Am.* **TuD7** June
[41] Lewis J P 1995 *Vision Interface 95* Canadian Image Processing and Pattern Recognition Society p 120
[42] Lo Y C, Cai J Y, Tasi M S, Tasi Z Y and Sun C C 2014 Side-illuminating LED luminaires with accurate projection in high uniformity and high optical utilization factor for large-area field illumination *Opt. Express* **22** A365–75
[43] DIALux (https://www.dial.de/en/dialux/)
[44] Xue Z 2017 Sidewall effect on the blue/white light pattern of sapphire-based LEDs *MS Dissertation* National Central University
[45] Sun C C *et al* 2014 Packaging efficiency in phosphor-converted white LEDs and its impact to the limit of luminous efficacy *J. Solid State Lighting* **1** 1–17
[46] Lin H J, Sun C C, Wu C S, Lee X H, Yang T H, Lin S K, Lin Y J and Yu Y W 2019 Design of a bicycle head lamp using an atypical white light-emitting diode with separate dies *Crystals* **9** 659
[47] Jen Y C 2016 The study of LED marine beacons *MSc Thesis* National Central University
[48] Xue Z 2017 Side-wall effect on the blue/white light pattern of sapphire-based LEDs *MSc Thesis* National Central University
[49] Li D R 2011 The study of package efficiency for white LEDs *MSc Thesis* National Central University
[50] Chen K Y 2018 Study of high-performance LED projection lamp for long-distance projection *MSc Thesis* National Central University
[51] Lin H J 2020 Optical design of bicycle head lamps based on K-mark regulation using a LED with separate dies *MSc Thesis* National Central University

Chapter 6

LED component-level secondary optics

LED secondary optics is based on the accurate light source model, designing an appropriate optical element to spread or project lights to the desired target. Although small in size, LEDs are still relatively complex compared to point sources. However, the optical elements available for LED secondary optics are highly diverse, coupled with design approaches covering both non-imaging and imaging optics. Therefore, some new composite optical elements have been developed. This chapter will begin with the principles of optical flux transfer, introduce all feasible optical elements, and discuss two typical optical applications: collimating and diffusing.

6.1 Essential principle of optical flux transfer

Secondary optics is an approach to collect the optical flux by the LED and redirect the flux to the illuminated target. No matter what the secondary optics is, the optical elements always obey the fundamental principle of optics. However, there are some essential rules of flux transfer through an optical system. Thus, before introducing secondary optics, we need to discuss the essential principle of optical flux transfer.

In an optical system, the optical flux is transferred from one plane to the other, as shown in figure 6.1. Through Hamilton's point characteristic function, we can describe the optical path from P to P' when passing the optical system with the expression [1]

$$n\hat{u} = -\nabla V, \tag{6.1}$$

$$n'\hat{u}' = \nabla' V, \tag{6.2}$$

where V is the optical path length from P to P'; n and n' are the refractive index of the space for P and P', respectively; u is the unit vector from P to P', $\hat{u}'$ is the unit vector from P' to P. The coordinates for P are (x, y, z), and for $P'(x', y')$ are (x', y', z').

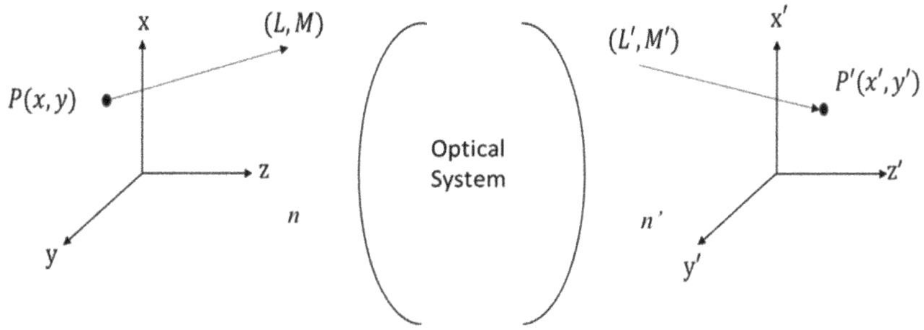

Figure 6.1. The optical system from the space of (x, y, z) to the space of (x', y', z').

Any ray emitted from P has the ray vector (L, M, N), where L, M, and N are the vector projections along the x, y, and z axes, respectively, and $L^2 + M^2 + N^2 = 1$. The ray passing through the optical system directs to P' with a vector projection (L', M'). If there is a small amount of displacement of $P(x, y)$, i.e., dx and dy, the corresponding vector projection variance is dL and dM. Thus (dL', dM') is the corresponding vector projection of the displacement amount of (dx', dy').

Through Hamilton's point characteristic function, the relation between (L, M, N) and (L', M', N') can be written

$$n\left(L\hat{i} + M\hat{j} + N\hat{k}\right) = -\left(\frac{\partial V}{\partial x}\hat{i} + \frac{\partial V}{\partial y}\hat{j} + \frac{\partial V}{\partial z}\hat{k}\right), \tag{6.3}$$

$$n'\left(L'\hat{i} + M'\hat{j} + N'\hat{k}\right) = \left(\frac{\partial V}{\partial x'}\hat{i} + \frac{\partial V}{\partial y'}\hat{j} + \frac{\partial V}{\partial z'}\hat{k}\right). \tag{6.4}$$

Equations (6.3) and (6.4) can be written

$$nL = -\frac{\partial V}{\partial x}, \tag{6.5}$$

$$nM = -\frac{\partial V}{\partial y}, \tag{6.6}$$

$$n'L' = \frac{\partial V}{\partial x'}, \tag{6.7}$$

$$n'M' = \frac{\partial V}{\partial y'}. \tag{6.8}$$

Through equations (6.3) and (6.4), we can obtain

$$\frac{\partial^2 V}{\partial x \partial x'}\frac{\partial^2 V}{\partial y \partial y'} - \frac{\partial^2 V}{\partial y \partial x'}\frac{\partial^2 V}{\partial x \partial y'} = \frac{\partial^2 V}{\partial x' \partial x}\frac{\partial^2 V}{\partial y' \partial y} - \frac{\partial^2 V}{\partial y' \partial x}\frac{\partial^2 V}{\partial x' \partial y}. \tag{6.9}$$

Using equations (6.5)–(6.8), equation (6.9) can be rewritten

$$n'^2\left(\frac{\partial L'}{\partial x}\frac{\partial M'}{\partial y} - \frac{\partial L'}{\partial y}\frac{\partial M'}{\partial x}\right) = n^2\left(\frac{\partial L}{\partial x'}\frac{\partial M}{\partial y'} - \frac{\partial L}{\partial y'}\frac{\partial M}{\partial x'}\right). \tag{6.10}$$

Using Jacobians transform, we can obtain

$$\frac{\partial L'}{\partial x}\frac{\partial M'}{\partial y} - \frac{\partial L'}{\partial y}\frac{\partial M'}{\partial x} = \frac{\partial(L', M')}{\partial(x, y)}, \tag{6.11}$$

$$\frac{\partial L}{\partial x'}\frac{\partial M}{\partial y'} - \frac{\partial L}{\partial y'}\frac{\partial M}{\partial x'} = \frac{\partial(L, M)}{\partial(x', y')}, \tag{6.12}$$

Equation (6.10) can be rewritten

$$n'^2\frac{\partial(L', M')}{\partial(x, y)} = n^2\frac{\partial(L, M)}{\partial(x', y')}. \tag{6.13}$$

Then equation (6.13) becomes

$$n'^2dx'dy'dL'dM' = n^2dxdydLdM, \tag{6.14}$$

where

$$dLdM = \frac{\partial(L, M)}{\partial(x', y')}dx'dy', \tag{6.15}$$

$$dL'dM' = \frac{\partial(L', M')}{\partial(x, y)}dxdy. \tag{6.16}$$

Equation (6.14) is called the generalized Lagrange invariant [1, 2]. In the polar coordinate system shown in figure 6.2, the ray vector projection can be written

$$(L, M, N) = (\sin\theta\cos\varphi,\ \sin\theta\sin\varphi,\ \cos\theta) \tag{6.17}$$

where θ is the angle on the yz plane, and ϕ is the angle on the xy plane. Then equation (6.15) is expressed

$$\begin{aligned} dLdM &= \frac{\partial(L, M)}{\partial(\theta, \varphi)}d\theta d\varphi \\ &= \left(\frac{\partial L}{\partial\theta}\frac{\partial M}{\partial\varphi} - \frac{\partial L}{\partial\varphi}\frac{\partial M}{\partial\theta}\right)d\theta d\varphi \\ &= \cos\theta\ \sin\theta d\theta d\varphi \\ &= \cos\theta d\Omega, \end{aligned} \tag{6.18}$$

where $d\Omega = \sin\theta d\theta d\phi$. Through the same procedure,

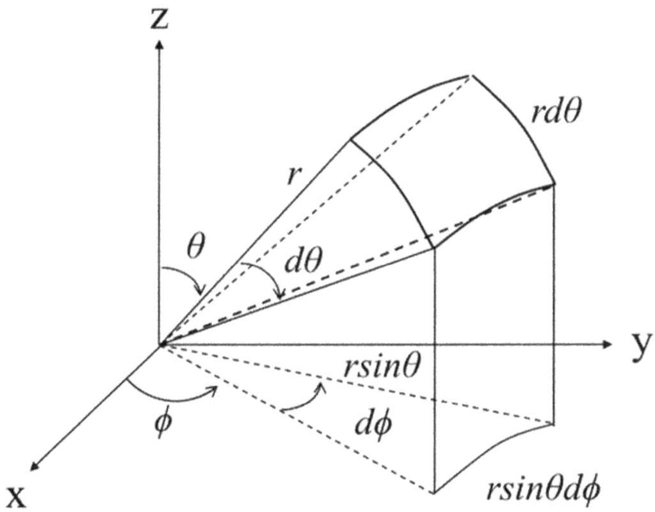

Figure 6.2. The polar coordinate system.

$$dL'dM' = \cos\theta' d\Omega, \tag{6.19}$$

where $d\Omega' = \sin\theta' d\theta' d\phi'$. With combining equations (6.18) and (6.19), equation (6.14) becomes

$$n^2 dA \cos\theta d\Omega = n'^2 dA' \cos\theta' d\Omega'. \tag{6.20}$$

Equation (6.20) is an expression of a unit of étendue, which is the product of the projected area of the light source and the solid angle of light emission. Here we denote dU as the étendue unit, and it can be written [3]

$$dU = n^2 dA \cos\theta d\Omega. \tag{6.21}$$

Now we can calculate the étendue of a Lambertian light source, where the cross-section area is A. The étendue of the Lambertian light source is written

$$U = \iint dU = n^2 A \iint \cos\theta d\Omega, \tag{6.22}$$

where $d\Omega = 2\pi \sin\theta d\theta$.

$$U = 2\pi n^2 A \int_{\theta=0}^{\frac{\pi}{2}} \cos\theta \sin\theta d\theta, \tag{6.23}$$

$$= \pi n^2 A. \tag{6.24}$$

In comparison with equations (6.14) to (6.17), for a Lambertian light source with a radiance of B, the total emission flux is written

$$F = BU/n^2. \tag{6.25}$$

$$dF = BdA\cos\theta d\Omega, \tag{6.26}$$

$$= \frac{B}{n^2}dU. \tag{6.27}$$

Etendue is a geometrical factor to check how the beam flow through different components, and it is also called throughput [3, 4], which is equal to the product of a cross-section area and the projection solid angle,

$$T_{1\to2} = A_1\cos\alpha_1\Omega_1\cos\alpha_2 = \frac{A_1A_2\cos\alpha_1\cos\alpha_2}{d^2}, \tag{6.28}$$

$$T_{2\to1} = A_2\cos\alpha_2\Omega_2\cos\alpha_1 = \frac{A_1A_2\cos\alpha_1\cos\alpha_2}{d^2}, \tag{6.29}$$

where α_1 and α_2 are the slanted angle versus the optical axis connecting the two cross-section areas, as shown in figure 6.3. Thus $T_{1\to2} = T_{2\to1} = T$, and the throughput is invariant. The invariant property is important in the optical design for a lighting system when applying a secondary optics. If there is an optical element, e.g., a lens, as used in the system shown in figure 6.4, and the full size of the lens and detector is utilized, it is applicable to write the invariance of étendue [3, 4]

$$U = T = A_e\Omega_{eo} = A_o\Omega_{oe} = A_o\Omega_{od} = A_d\Omega_{do}, \tag{6.30}$$

where A_e, A_o and A_d are the area of the emitter, lens and detector, respectively; Ω_{eo}, Ω_{oe}, Ω_{od}, and Ω_{do} are the solid angle of the emitter to the lens, lens to the emitter, lens to the detector, and detector to the lens, respectively.

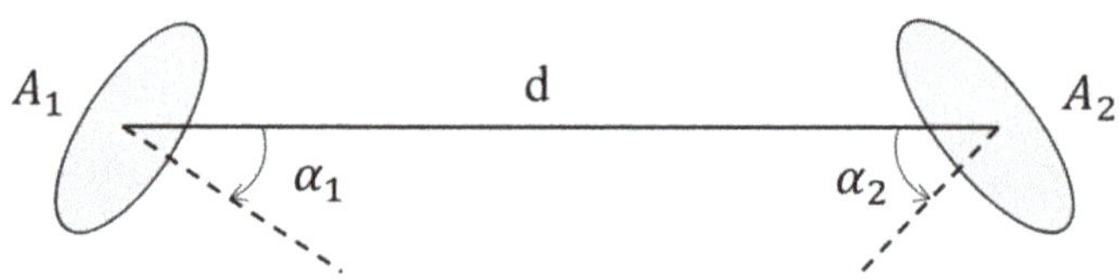

Figure 6.3. The throughput invariant between two cross-section areas.

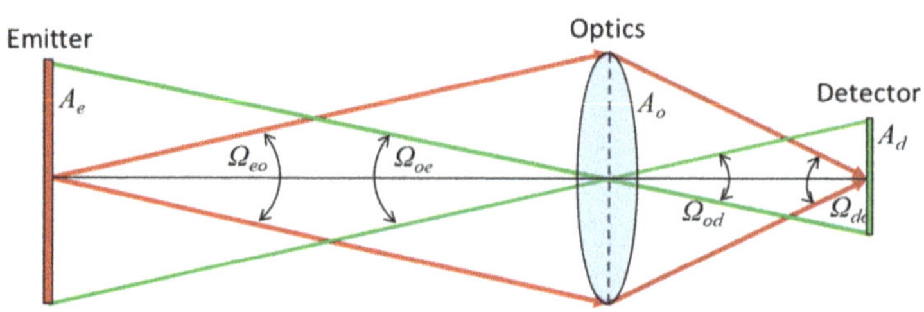

Figure 6.4. Schematic diagram of the optical flux from an emitter to a detector with full area utilization performing throughput invariant.

6.2 Typical optical elements for secondary optics

Secondary optics is aimed at projecting light emitted by an LED to the target with an expected illumination pattern. There are various elements available for the purpose. The basic elements include lenses (positive or negative, bulk or slim type), mirrors (single or multiple segments), light tubes (hollow or solid), diffusers (surface or volume scattering), and gratings or diffractive optical elements. All the elements are in different forms or sizes with different functions.

Generally, a lens or a mirror has a focal power, and thus such an element is applicable to optical imaging. In general, an illumination pattern is formed by non-imaging approaches rather than imaging approaches. However, an imaging element could be more useful in some specific applications. Figure 6.5 shows a positive and a negative lens. A positive (negative) lens is able to bend the incoming light toward (outward) to the optical axis. If a collimating light is incident on a lens, almost all the rays will pass through the focal point and form a focusing spot. Ideally, all the rays will be focused at the focal point so that the focusing spot is bright and small, but actually the focusing is not perfect. Some of the rays do not pass through the focal point when the lens has aberrations or the alignment is incorrect. A lens is said to be diffraction-limited when there is no aberration. Even so, the focusing spot at the focal point is not a point. In fact, the size of the focused spot is limited and is decided by diffraction, which originates from the inherent property of an optical wave.

In some cases, the thickness of a lens is limited. Thus a slim type of lens is demanded, and the most common slim lens is called Fresnel lens [5–7], as shown in figure 6.6. Typically, a Fresnel lens is a lens in a plate type. The forming of a Fresnel lens is by reducing the thickness of a general lens by dividing several radial segments

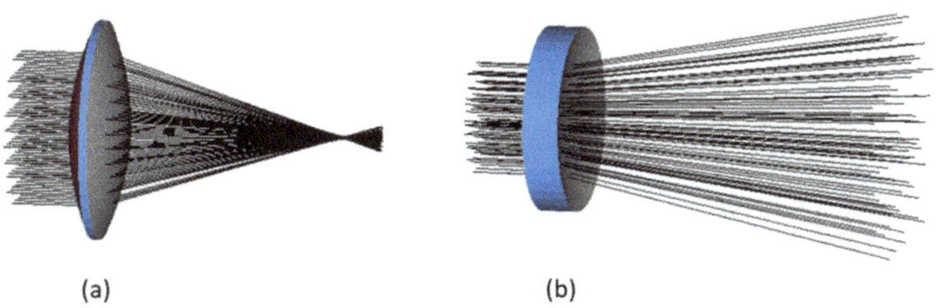

Figure 6.5. (a) A positive (convex) lens, and (b) a negative (concave) lens.

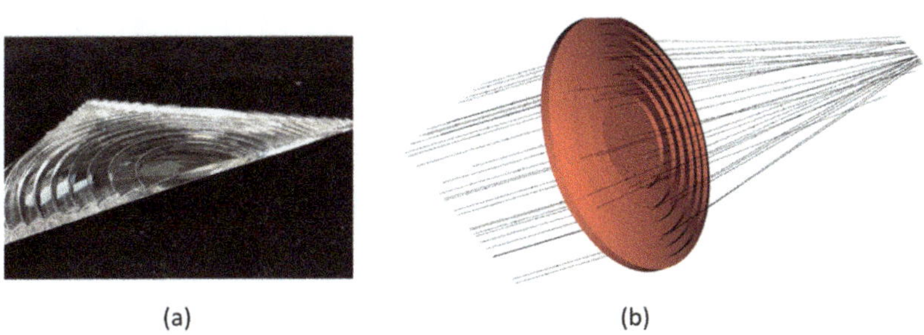

Figure 6.6. (a) Cross-section of a Fresnel lens, and (b) a positive Fresnel lens is used for focusing.

across the horizontal plane. Unfortunately, this induces additional round angles and surface gaps on the boundary between two segments. Both the round angle and surface gap will perform unwanted reflection or refraction to form stray light and reduce energy efficiency.

In addition to a lens, a mirror is another element well workable to imaging function. A concave mirror has a positive power and a convex lens has a negative power. The main mechanism of a mirror is reflection, as shown in figure 6.7. The reflection efficiency or reflectivity is one of the key factors of a mirror. Typically, the surface of a mirror is coated with metal film or dielectric film to increase reflectivity to as high as 90% or above. Dielectric film is more selective in wavelength than metal film is, and a very high reflectivity is achievable through multiple layers. In comparison with a lens, one of the advantages of a mirror is that it is free of dispersion, which is an inherent property of an optical medium, where the refractive index is a function of wavelength. Using a mirror rather than a lens it is possible to tighten the system thickness. However, a mirror could be more sensitive in alignment than a lens.

Light tubes are pipe structures which are able to direct the incoming light to the other end, as shown in figure 6.8. That a light tube can guide light propagation is called a light guide, which could be hollow or solid. A hollow light guide must have a high-reflective inner surface to avoid serious absorption so that most energy can be

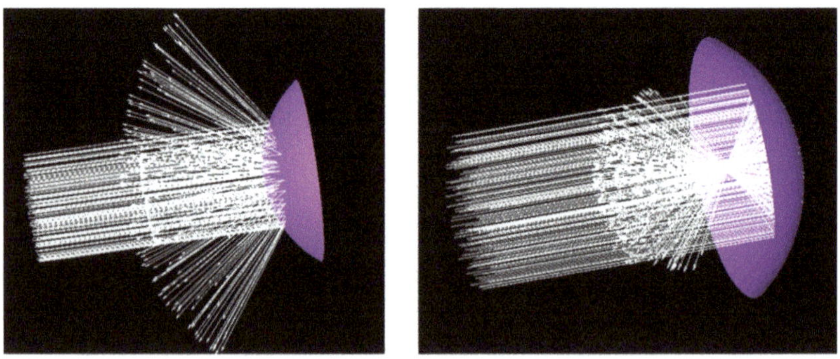

(a) (b)

Figure 6.7. (a) A negative (convex) mirror, and (b) a positive (concave) mirror.

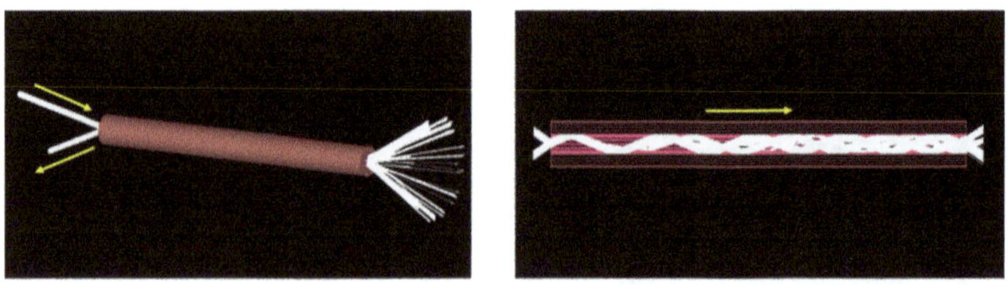

(a) (b)

Figure 6.8. (a) A light tube can guide an incoming light to another end, and (b) the optical path in a light guide.

kept when reaching the other end of the light guide. A solid light guide is similar to a an optical fiber, where the incoming light at an angle smaller than the acceptance angle of the fiber or within the numerical aperture (NA) will successfully encounter multiple times of total internal reflection, so that the incoming light can travel along the light guide with less energy loss. A fiber bundle is capable of transmitting optical flux and optical image as well, as shown in figure 6.9.

In imaging optics, all the surfaces of the optical elements are smooth. But in illumination, whether the surface is smooth or rough is decided by the optical design. In a rough surface, the reflected (transmitted) light seems not to follow the reflection (refraction) law if the surface is regarded as a planar surface. Actually, there are a lot of tiny slanted reflection (transmission) surfaces with unexpected angles across the surface to cause the unpredicted reflection (transmission). Thus, each reflected (transmitted) light actually follows the reflection (refraction) law at each tiny surface. Due to the characteristic of the reflected (transmitted) light, we could use 'scattering' to illustrate the phenomena, as shown in figure 6.10(a). In contrast to a

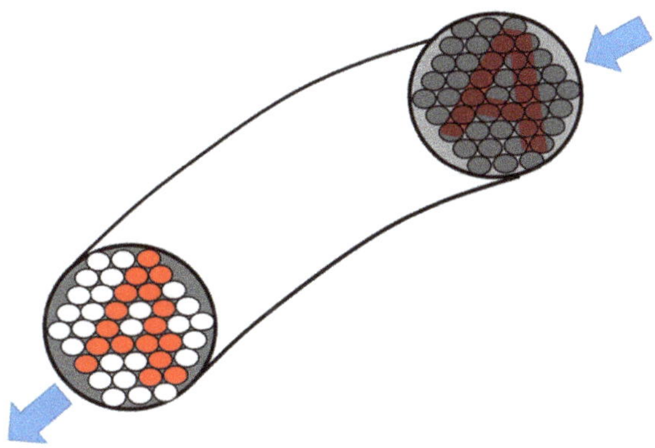

Figure 6.9. A fiber bundle can be used to transmit an image.

Figure 6.10. (a) The rough surface will cause scattering phenomena. (b) Schematic diagram of a volume diffuser.

rough surface, a diffuser can have smooth surfaces but with scattering particles inside the volume of the diffuser, and such an element is called volume diffuser, as shown in figure 6.10(b). Those scattering particles can be with high reflectivity or with different refractive indices. The former is to act tiny reflectors inside the volume and the latter is to act tiny optical boundaries in the volume, and both of them can change propagation direction of the incoming light, and cause scattering.

Design of the secondary optics is an art to meet the expected illumination requirement on the target with the most appropriate optical elements, which are not just by a single optical function. For example, if one tries to collect the most flux by an LED and collimate the light, there are two ways, i.e., using a lens or a mirror. However, if a lens is used, the LED should be put at the front focus of the lens, and the light is collimated. But the efficiency is low because the rays at larger angles could not be collected by the lens. The condition is similar when a concave mirror is used. Therefore, a TIR (total internal reflection) lens is proposed for this purpose, as shown in figure 6.11. TIR means that the collected rays encounter total internal reflection in the lens volume, which is especially useful to the rays at larger angles. Along the normal direction of a TIR lens, there is a thick lens to collect the rays at smaller angles. As a result, almost all the rays emitted by an LED can be collected. A TIR lens is a solid lens with a specific shape. Typically, the lens is made with the use of a plastic medium through injection molding. Deformation in the injection process and the heavy weight are two negative factors in designing a TIR lens. Generally the TIR lens is kept small. An alternative with higher design freedom is a freeform lens [8], where two surfaces, including the entrance and the exit surface, are designed according to ray vectors from the light source and the illumination target. Usually, a point source rather than the real light source is used to design the initial surfaces and modification will be made when the real light source is applied.

Another useful design for the secondary optics is to use a multi-segment reflector to direct the incoming light to the illumination target, as shown in figure 6.12. Each segment of the reflector has its specific curvature and orientation to redirect the light. Such a complex optical element is frequently used in automotive headlamps, which needs to meet a regulation with a strict cutoff line so that all the segments need to be well designed and fabricated [9].

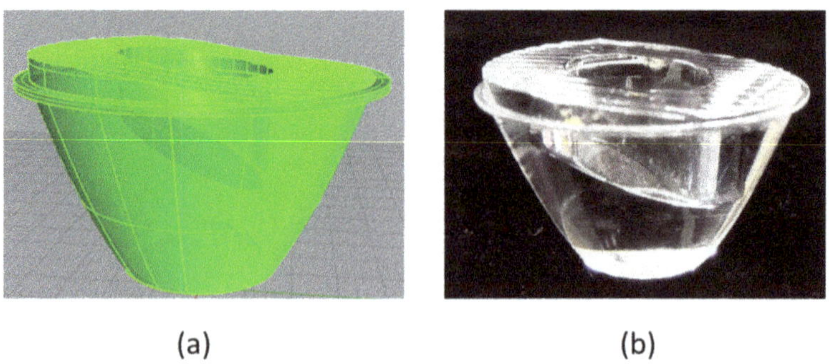

Figure 6.11. (a) A structure of a TIR lens, and (b) the corresponding prototype.

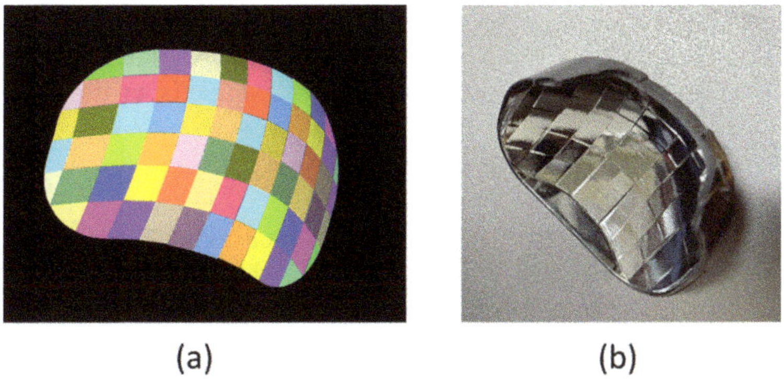

Figure 6.12. (a) The structure of a multi-segment reflector, and (b) the prototype.

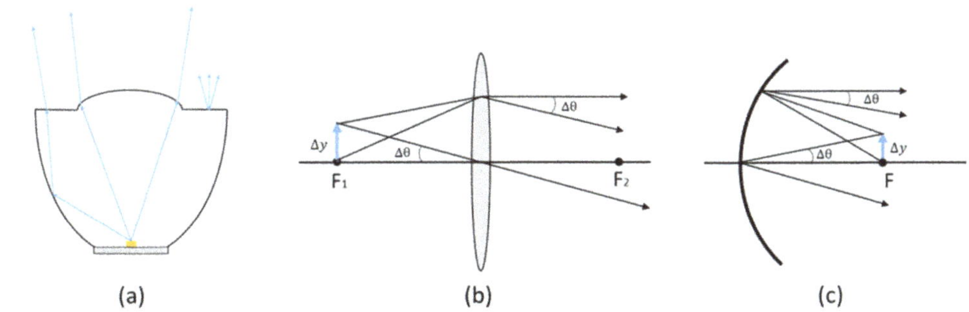

Figure 6.13. (a) A schematic diagram of a TIR lens. The lateral deviation will cause angular deviation of (b) a lens, and (c) a reflector.

6.3 High-directional LED illumination

In comparison with the other light sources, LED is rather smaller, and is a semiconductor component. The inherent characteristic of an LED light source is to form a directional light source through a secondary optical design. In general, a directional light source should be in high intensity, i.e., high optical flux in a unit solid angle. However, the compactness of an LED does not provide this advantage because the optical exitance of an LED in the typical size is smaller than the other traditional light sources, such as the high-intensity discharge (HID) lamp. Therefore, LED light sources are suitable to project to a distance as long as hundreds of meters rather than kilometers. If LEDs are used for long-range projection, LED clusters are needed, and the whole size will be large.

Projection of the LED light to long distance is a complicated problem, which contains various considerations. The main problem is how to collimate the LED light in a limited volume and acceptable power. If compactness is the point, a TIR lens is the best solution and is applicable. The TIR lens contains two parts, one is a positive lens along the optical axis, and the other is a reflector performing total internal reflection. As shown in figure 6.13, to see the divergent angle of a TIR lens, we can calculate the spread angles by the lens and the reflector through Gaussian geometrical optics. Figure 6.13(b) is for a lens. If there is a lateral extension of the LED die, Δy shown in figure 6.13(b), the divergent angle is written

$$\Delta\theta = \tan^{-1}\frac{\Delta y}{f}, \tag{6.31}$$

where f is the focal length of the lens. Similarly, equation (6.31) is applicable to the reflector, as shown in figure 6.13(c). Therefore, the essential problem to cause angular spreading is the lateral size of the LED. Please note that the focal length of the lens could be different from that of the reflector.

If one of the elements (the lens or the reflector) is responsible for collecting more flux by the LED light source, the intensity of the TIR lens will be decided by the reflector, but the divergent angle will be decided by the element with the shorter focal length. For example, a solid TIR lens is the size of 3 cm × 3 cm × 3 cm, shown in figure 6.14, where the diameter of the lens on the top surface is around 7 mm. A 2 mm × 2 mm Lambertian emitter is used to replace an LED light source, and is immersed in the solid TIR lens. The Lambertian emitter is located at the common focus of the lens and the reflector of the TIR lens, but the focal length of the lens and the reflector is 30 mm and 1.77 mm, respectively. In the simulation, the total flux by the Lambertian emitter is 100 lm, the simulated intensity is 2511 cd, where 2376 cd is by the reflector, and 135 cd is by the lens. It is clear to see that the reflector collects more flux than the lens, so the intensity is decided by the reflector as prediction. Also, since the focal length of the lens is longer, the divergent angle is decided by the lens, and is simulated 2.7°, which is the same as that by equation (6.31).

If the divergent angle is requested to be further small, the only solution is to enlarge the focal length of the lens and the reflector as well while the size of the LED light source is kept unchanged. To collect most flux by the light source, the extension of the focal length means that the dimensions of the reflector and the lens should be larger. Such a situation will make a TIR lens more impractical owing to heavy weight. A big-size optical element should not be a single solid element like a TIR lens. Therefore a good choice is to divide a TIR lens into two parts, one is an independent lens and the other is a collimating mirror. Figure 6.15 shows a schematic diagram of an optical system with compound elements, where a lens is mounted on a parabolic mirror. If the focal length of the optical elements were twice that of the TIR lens, the divergent angle could be reduced to half of the TIR lens.

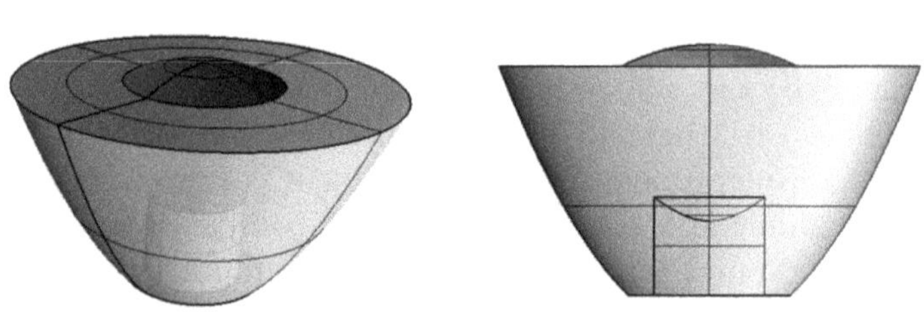

Figure 6.14. The geometry of a typical TIR lens.

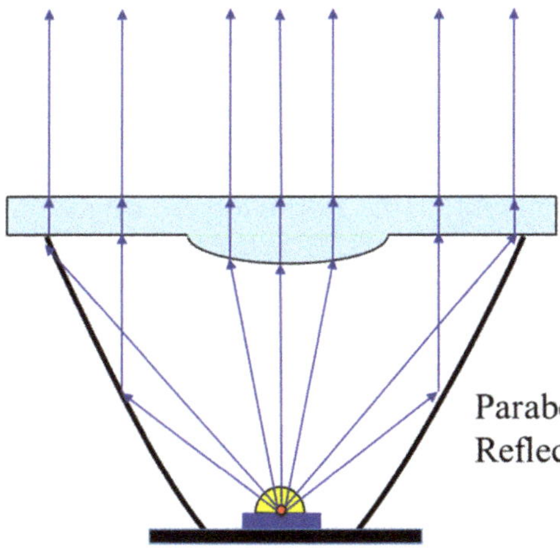

Figure 6.15. A compound optical element for well collimation [10].

To illuminate a target at a much longer distance, e.g., 10 km, a cluster of LED light modules could be put together to reach a high-intense spot. According to ANSI regulation, the projection distance is defined when illuminance at the target is 0.25 lux. To reach the goal of the distance 10 Km, the total optical flux and the requested intensity must be figured out. A useful factor, the optical utilization factor (*OUF*) is important for estimation. *OUF* is defined as the ratio between the flux at the target and the flux by the light source. If the area of the illumination target is known, the required total flux can be calculated through the *OUF*. To simplify the calculation, if the target area under illumination is A, the divergent angle is

$$\theta = \frac{\frac{\sqrt{A}}{2}}{d}, \tag{6.32}$$

where d is the projection distance. If the lateral dimension of the LED is t, the focal length of the major element of the projector is written

$$f = \frac{td}{\sqrt{A}}. \tag{6.33}$$

For example, if t = 0.5 mm, and d = 10 000 m, the lateral dimension of the illuminated area is 100 m, and then the focal length is calculated 5 cm, which could determine the possible size of the optical system. Once the geometrical size is determined, through the *OUF*, the total flux (F) needed for the LED cluster is estimated

$$F = \frac{A}{4(OUF)}. \tag{6.34}$$

Figure 6.16 shows a design following the principle illustrated through equations (6.31)–(6.34). In total, 144 pcs of pcW-LED are put together. The optical system of compound elements is to project light by a pcW-LED. Through precise calculation, the diameter of the lens is at set 32 mm, and the diameter of the top opening of the reflector is set 65 mm, while the height of the reflector is set 40 mm. The simulation shows that the central illuminance (the full divergent angles) are around 2515 lux (1.38°), 1,034 lux (1.2°) and 3549 lux (1.33°), for the parabolic reflector, the cover lens and the whole system, respectively, when a stable flux of 340 lm under a designed heat dissipation structure with power injection around 4.3 W is determined. Figure 6.17 shows the system where nine modules are contained. The total power of each module is 68.2 W, the injection current is 1.25 A, the divergent angle is 1.6°, the maximum luminous intensity is 2 840 000 cd, which could reach 3.37 Km according to ANSI regulation. The whole system is with dimensions of 1.4 m × 1.4 m, the total power consumption is 615 W, and the weight is around 50 kg. Such an LED lighting system for long-range projection presents the inherent characteristic of a

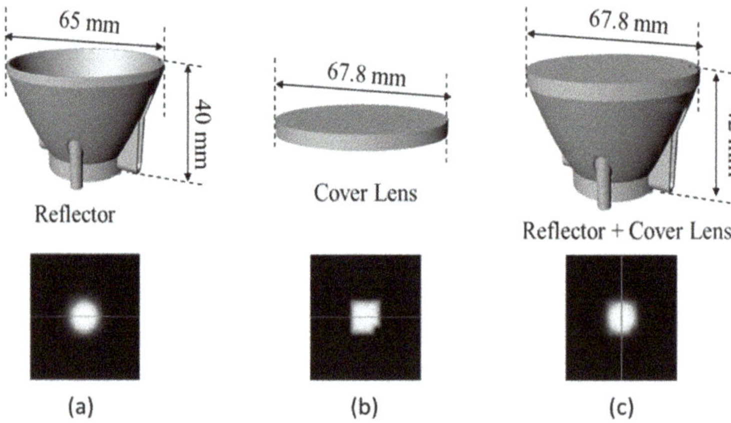

Figure 6.16. (a) The reflector of the compound element, and the projection spot, (b) the lens and the projection spot, and (c) the compound element and the projection spot. Reprinted with permission from [14].

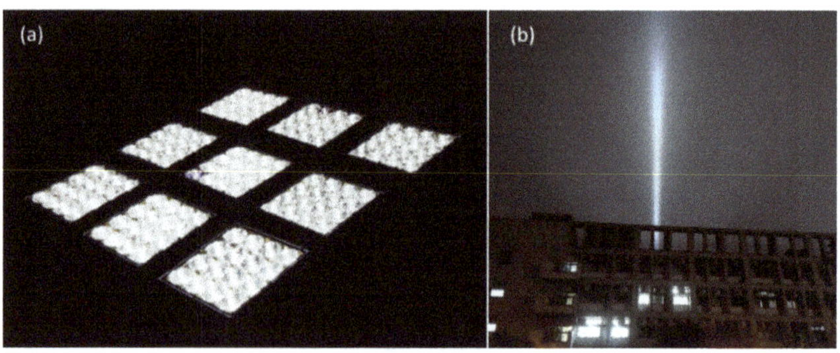

Figure 6.17. (a) The whole system contains nine modules, (b) the projection spot watched from a far place [10].

semiconductor light source. The injection power is relatively small, the weight is acceptable, but the total volume or size for an LED high-intense projector is relatively larger than a traditional light source for this purpose.

6.4 Angular radiation extension by a diffuser

One of the advantages of LED light sources is compactness. The typical emitting area of a high-power LED is around 1 mm × 1 mm, where the injection power is around 3–5 W. If a luminaire requires more power injection, one of the approaches is to add more LEDs in the luminaire. However, even if the provided optical flux is large enough, the optical property of the light source should be managed. One of the most important issues is anti-glare and uniformity on the exit face. One of the approaches is to use a diffuser to reduce the glare induced by the LED array, to enlarge the divergent angle and to reduce the luminance on the exit face.

There are three types of diffuser, including surface texture, volume scattering and microlens array. The first type is called surface-textured diffuser (simplified to ST-diffuser), where the surface is textured with random structures, as shown in figure 6.18. An ST-diffuser can enlarge the divergent angle of an incoming light to a certain degree owing to refraction on roughened surfaces. Owing to random textured surfaces, the refraction angle of each light on the surface is not predictable. The angle spreading relates to structure depth and average period of the texture pattern. The larger depth and smaller period could induce larger divergent angles. Once the divergent angle is enlarged, the intensity and the luminance of the luminaire could be reduced.

The second type of diffuser is called volume scattering diffuser (simplified VS-diffuser). As shown in figure 6.19, a VS-diffuser contains scattering particles inside the diffuser. In general, the scattering particle is SiO_2 in a round shape. Rather than the other materials, SiO_2 exists in many crystalline forms, called polymorphs. The different crystal form will have different density and refractive index as well. Figure 6.20 shows the refractive index as a function of density. The refractive index difference between the SiO_2 particle and the plastic sheet creates optical boundaries in a VS-diffuser, and this is the scattering mechanism of a VS-diffuser. The angular

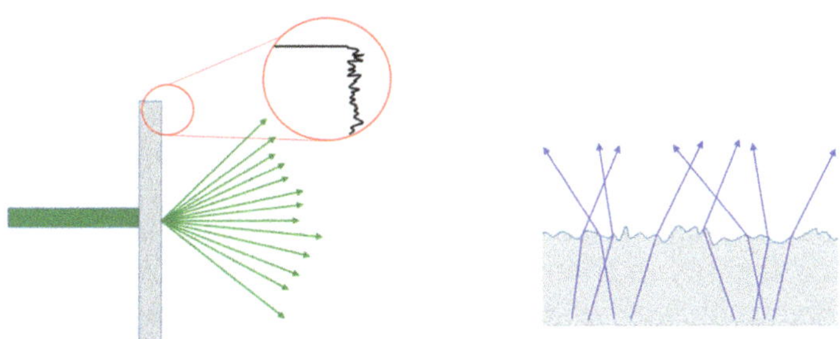

(a) (b)

Figure 6.18. (a) Surface texture in an ST-diffuser, and (b) the enlarged divergent angle owing to refraction on the textured surface.

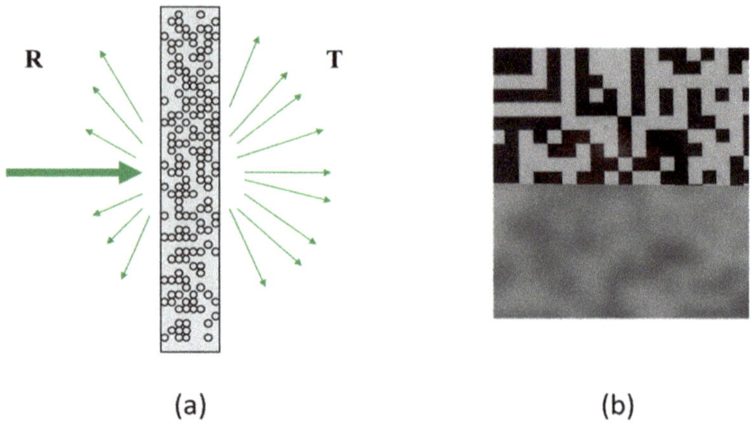

Figure 6.19. (a) Numerous SiO_2 particles inside a plate form a VS-diffuser. (b) The visual effect by a VS-diffuser.

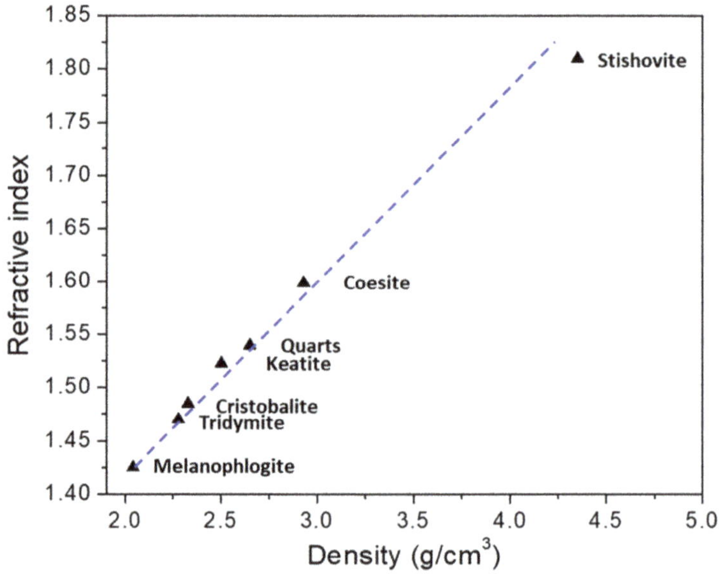

Figure 6.20. The refractive index as a function of density of SiO_2 [11].

extension function depends on how many times of scattering there are for an incoming light inside a VS-diffuser. Therefore, more SiO_2 particles inside a unit volume will induce more scatterings, and thus the divergent angle will be further enlarged. This is the way to control the angular extension function of a VS-diffuser.

If a larger divergent angle of a luminaire is required in an optical design, using a VS-diffuser is a good choice. However, larger angular extension will cause more backward scattering and decrease the transmission efficiency of a VS-diffuser. This is the inherent characteristic of a VS-diffuser. To conquer this problem, especially in LED lighting, using photon recycling is a useful approach. A luminaire of cuboid

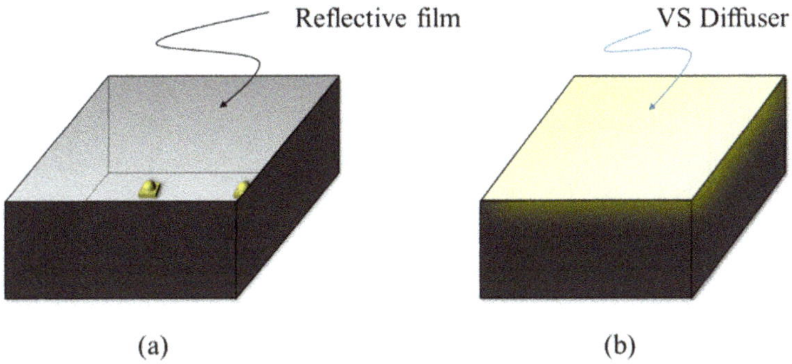

Figure 6.21. The schematic diagram of a luminaire of cuboid form covered with a VS-diffuser for photon recycling.

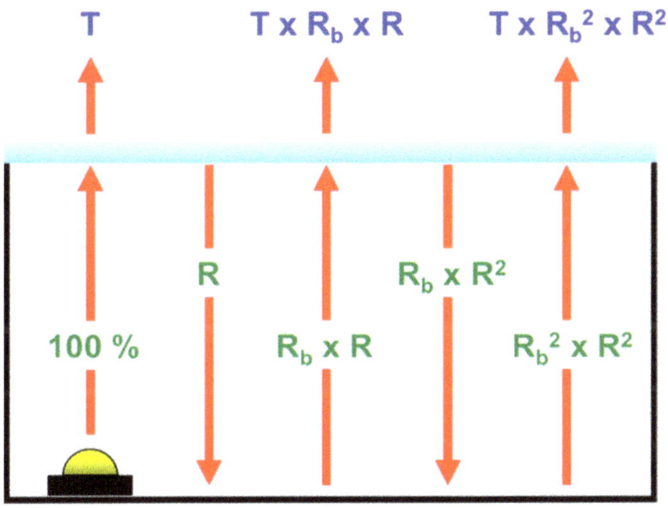

Figure 6.22. Photon recycling mechanism.

form covered with a VS-diffuser is used as an example, shown in figure 6.21. The VS-diffuser is to prevent one from viewing four discrete light sources and to reduce the luminance of the exit face. In the following example, there are three kinds of VS-diffuser, where the SiO_2 particle density is different. The largest density performs more angular extension, but the one-shot transmittance is around 55%. The transmittance of the other two diffusers are 60% and 70%, respectively. If the absorption of the VS-diffusers is neglected, the backward light bears the energy of 45%, 40% and 30% of the incoming light, respectively. Fortunately, the backward light can be recycled in a luminaire. If the other five surfaces inside the luminaire are coated with high reflective material, as shown in figure 6.22, the recycled light can have a good chance to escape from the cuboid. Assume the one-shot transmittance and reflectance are T and R, respectively, and the reflectance of the side wall is R_b, the cavity transmittance can be written [12]

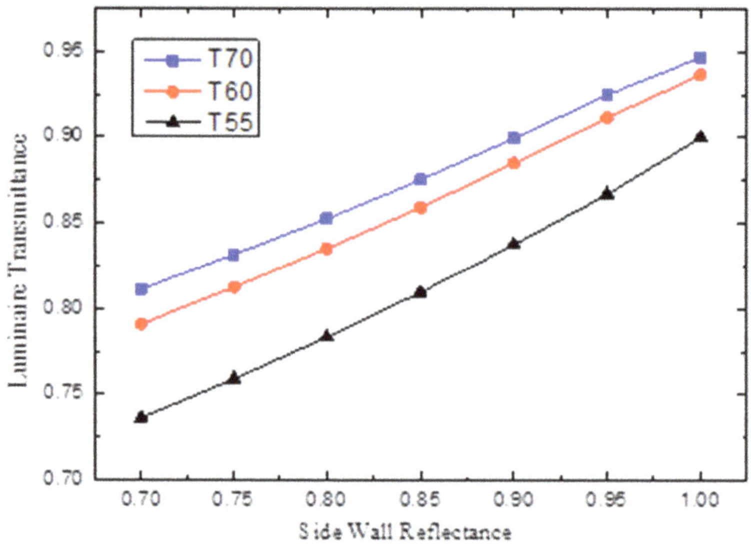

Figure 6.23. Luminaire transmittance versus side wall reflectance [12].

$$T_C = T + TR_bR + TR_b^2R + \ldots = \frac{T}{1 - RR_b} \tag{6.35}$$

Figure 6.23 shows the calculation according to equation (6.35). If R_b is near 100%, the cavity transmittances are around 95% for $R = 70\%$, 94% for $R = 60\%$ and 89% for $R = 55\%$, respectively. Practically, $R_b = 90\%$ is achievable, and then the cavity transmittances are 90% for $R = 70\%$, 88% for $R = 60\%$ and 83% for $R = 55\%$, respectively. The cavity transmittances are so amazing because photon recycling saves more than 20% of energy and the luminaire becomes homogeneous. At the same time, the luminaire reduces glare effect.

The third approach is to introduce a well-defined microlens array on a surface of a diffuser which is called a microlens diffuser (ML-diffuser) [13]. In contrast to an ST-diffuser, the surface structure is under special design. The structure could be a microlens array or other similar structures. Figure 6.24 shows an example. The shape of the microlens can be the same or different, depending on the demanded illumination pattern. If the illumination pattern is a flat rectangle, and the incoming light is collimating, the microlens could be like a cylindrical lens, and it means that the optical power along one axis is larger than that along the other axis. The shape of a microlens can be designed through paraxial geometrical optics, and a simulation through Monte Carlo ray tracing is helpful to figure out the light pattern. One of the most important issues to make an ML-diffuser is machining. Fabrication of a microlens structure has a limit on the lens size. Tightening the size could be very expensive in manufacture. An alternative way of iterative down-size molding process with glass sintering technology is applicable. Such a technology applies inherent shrinkage of a specific silica glass in the sintering process to form a downsized structure. Each sintering process could induce one-dimensional shrinkage to one half

Figure 6.24. The structure of an ML-diffuser.

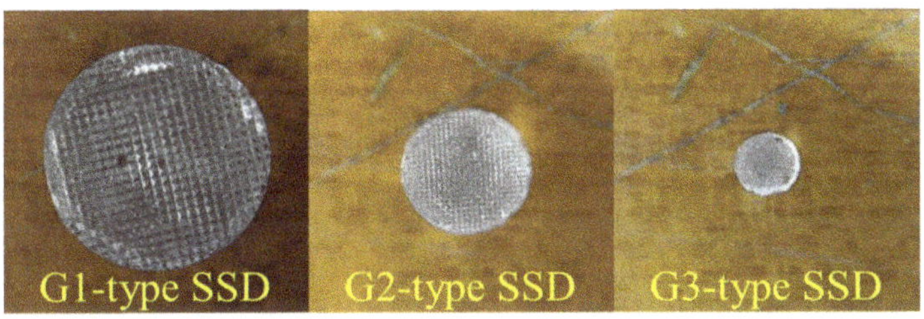

Figure 6.25. The three-generation molding samples in each sintering process.

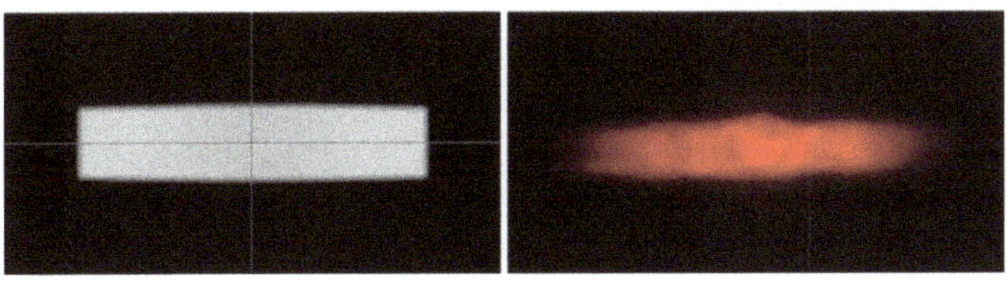

Figure 6.26. The simulated illumination patterns in flat rectangle shape from small (left) to large (right). Reprinted with permission from [15].

and one eighth in the volume. Thus, if three iterative sintering processes are applied, the machining structure can be eight times larger than the final one in one dimension. Figure 6.25 shows the molding samples in each sintering process. The observed light patterns with those ML-diffusers are shown in figure 6.26, where the illumination light is a red laser with small lateral beam width so that the light pattern does not look uniform.

References

[1] Lakshminarayanan V, Ghatak A K and Thyagarajan K 2002 *Lagrangian Optics* (Berlin: Springer)

[2] Mahajan V N 1998 *Optical Imaging and Aberrations: Ray Geometrical Optics* (Bellingham, WA: SPIE Press)

[3] Koshel R J 2013 *Illumination Engineering: Design with Nonimaging Optics* (New York: Wiley)

[4] Chaves J 2016 *Introduction to Nonimaging Optics* 2nd edn (Boca Raton, FL: CRC Press)

[5] Leutz R and Suzuki A 2001 *Nonimaging Fresnel Lenses: Design and Performance of Solar Concentrators* (Berlin: Springer)

[6] Davis A and Kühnlenz F 2007 Optical design using Fresnel lenses: basic principles and some practical examples *Opt. Photon.* **2** 52–5

[7] Xie W T, Dai Y J, Wang R Z and Sumathy K 2011 Concentrated solar energy applications using Fresnel lenses: a review *Renew. Sust. Energ. Rev.* **15** 2588–606

[8] Winston R, Miñano J C and Benitez P G 2005 *Nonimaging Optics* (Amsterdam: Elsevier)

[9] Sun C C, Wu C S, Lin Y S, Lin Y J, Hsieh C Y, Lin S K, Yang T H and Yu Y W 2021 Review of optical design for vehicle forward lighting based on white LEDs *Opt. Eng.* **60** 091501

[10] Wu C S, Chen K Y, Lee X H, Lin S K, Sun C C, Cai J Y, Yang T H and Yu Y W 2019 Design of LED spot light system with projection distance reaching 10 km *Crystals* **9** 524

[11] Skinner B J and Appleman D E 1963 Melanophilogite, a cubic polymorph of silica *Am. Mineral.* **48** 854–67

[12] Sun C C, Chien W T, Moreno I, Hsieh C T, Lin M C and Hsiao S L 2010 Calculating model of light transmission efficiency of diffusers attached to a lighting cavity *Opt. Express* **18** 6137

[13] Lee X H, Tsai J L, Ma S H and Sun C C 2012 Surface-structured diffuser by iterative down-size molding with glass sintering technology *Opt. Express* **20** 6135

[14] Chen K Y 2018 Study of high-performance LED projection lamp for long-distance projection *MSc Thesis* National Central University

[15] Lee X H 2012 Design and manufacture of surface-structured diffuser and the applications *PhD Thesis* National Central University

Chapter 7

LED system-level LED lighting optics

In Illumination design, the optical system design focuses on creating a comfortable lighting environment and achieving the best optical performance. The poor optical design may make what we see a less- or over-bright visual environment and even raise concerns about wasted light and light pollution. An ideal optical system design is first to maximize the optical utilization factor so that most of the light hits the target. The second is to use additional optics to block unwanted light to reduce unnecessary glare and light pollution. This chapter will introduce several lighting applications based on the system level. The optical design in these lighting systems will be discussed.

7.1 LED street/roadway lighting

White LED is a light source suitable for streetlighting because of its compactness, high efficiency, high color rendering index (CRI) and adjustable light pattern. Unlike the traditional light sources such as low-pressure sodium, an LED streetlight usually contains an LED array and the light exit face is flat unlike the others. The advantage is that less sky glow is induced, and will be helpful for dark sky or light pollution reduction.

LED streetlights can be divided into three main design configurations, including cluster LEDs with a single lens (simplified as CSL), an LED array accompanied by a lens array (simplified as ALA), and a tilted LED array with a reflector (simplified as TAR), as shown in figure 7.1 [1].

CSL is special for a white LED cluster, where the LED die array is covered with a phosphor layer. Thus the white LED is relatively big in comparison with a white LED with a single die. Such a white LED cluster is usually operated at a high-power level so that the heat dissipation is critical. Owing to higher temperature across the top surface, the lens is made with glass rather than plastic medium. The lens is then heavier. In order to spread the light pattern along the roadway, the optical design of the lens is special. A peanut-shape lens is proposed to perform an illumination

doi:10.1088/978-0-7503-2368-0ch7

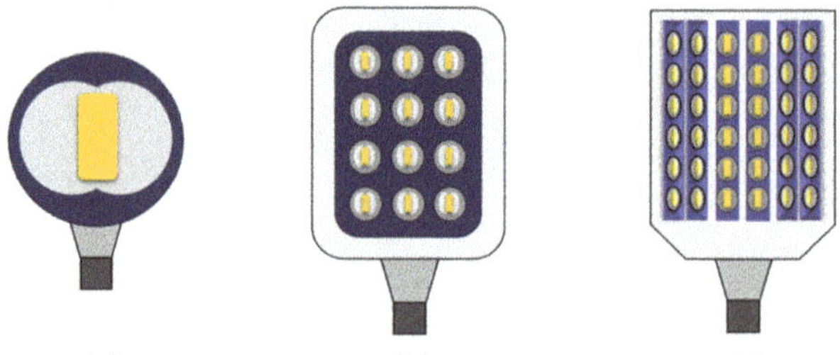

Figure 7.1. The three kinds of LED streetlight: (a) the CSL type, (b) the ALA type, and (c) the TAR type.

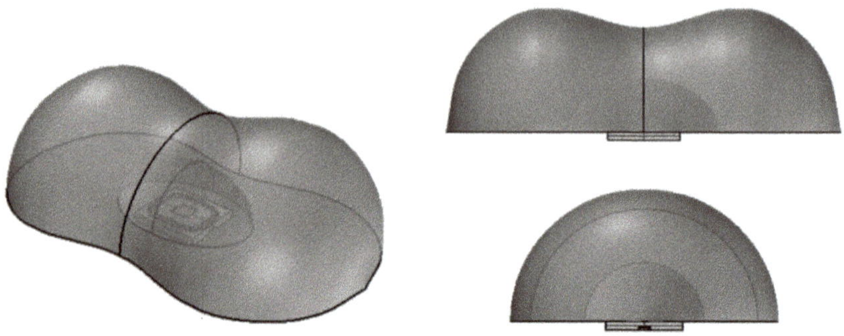

Figure 7.2. The geometry of the peanut lens.

pattern of a flat rectangle [2]. The peanut lens contains two main positive lenses on two sides, as shown in figure 7.2. The white LED is placed at the central position of the lens, so the light source is eccentric to one of the lenses along a lateral direction, and is eccentric to the other lens along the opposite direction. As a result, the light pattern on the roadway is a flat rectangle pattern, as shown in figure 7.3. Such a design offers a fitting pattern to a roadway, and it is an advantage of white LEDs. Since an LED cluster is big, so it is the cover lens, and it is not a lightweight design. In a certain application, two cluster LEDs are needed to provide sufficient illumination. Thus an alternative design called butterfly lens to combine two peanut lenses is proposed, as shown in figure 7.4 [3].

The challenges of heavy weight, heat dissipation and lifetime make a white LED cluster not widely acceptable in LED streetlight. To conquer these problems, among one of the solutions, ALA distributes LED arrays across the luminaire. Each discrete white LED is relatively small, and the cover lens is also small, so that the weight of the LED luminaire can be reduced. Also, the heat generated by the LED will be more uniformly spreading across the MCPCB board, as shown in figure 7.5. Besides, the surface temperature is relatively low for a single white LED operated at around 3–10 watt, so the cover lens array can be made with a single plastic plate. The geometry of the cover lens in the LED array is similar to that in the case of the white LED cluster. The power management of ALA is easy. For example, if the total power of an ALA luminaire is requested to be 100 W, and each single LED is designed to operate at 5 W, there are 20 pieces of white LED and lens in the

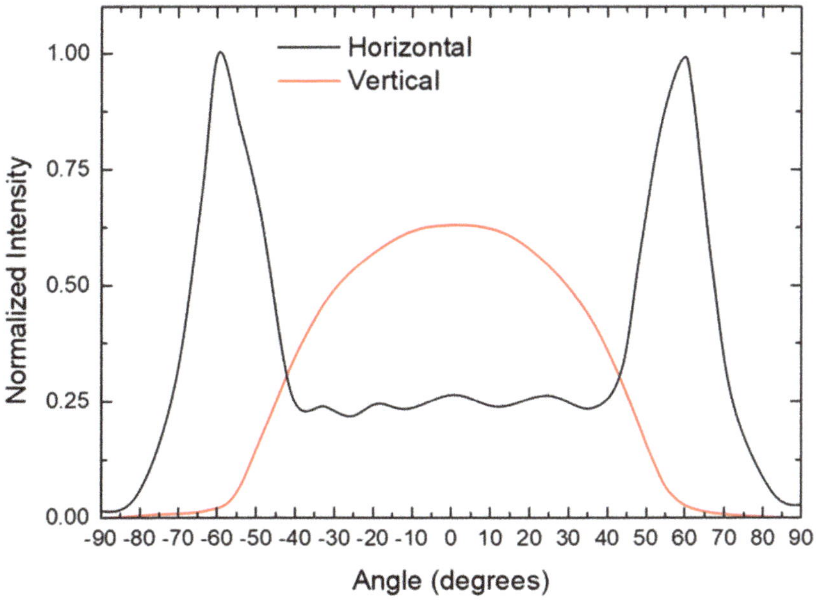

Figure 7.3. The angular distribution in the vertical and horizontal directions.

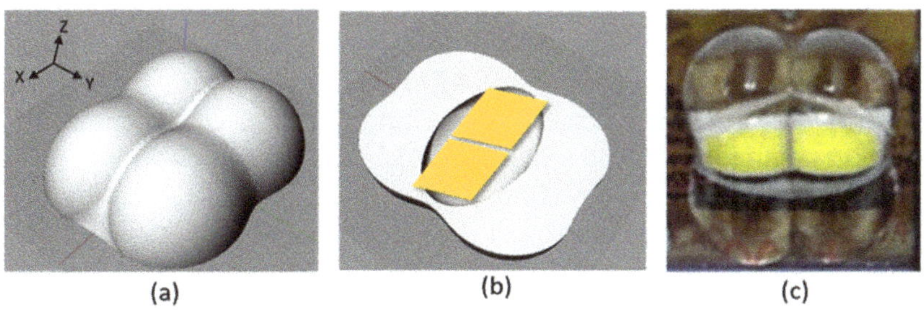

Figure 7.4. The geometry of the butterfly lens: (a) top view, (b) bottom view, (c) real photo.

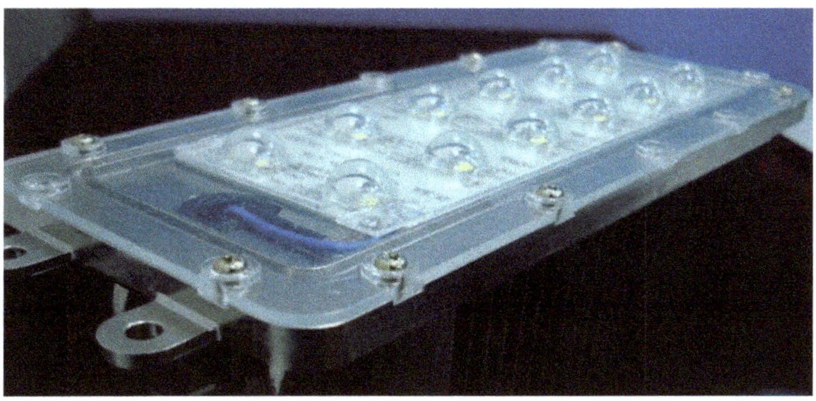

Figure 7.5. A photo of the ALA luminaire.

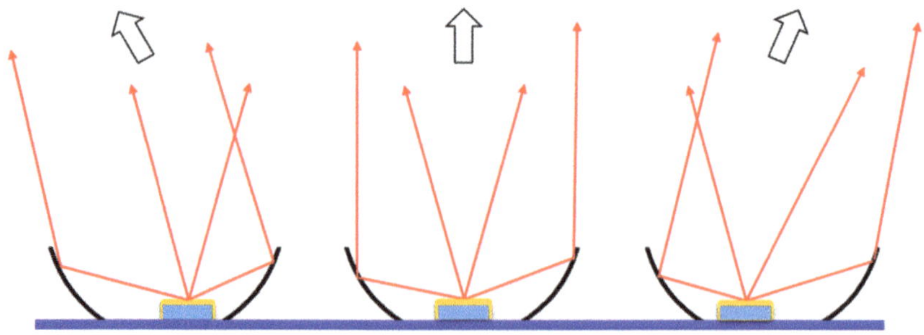

Figure 7.6. The design principle to spread the light pattern along the roadway for the TAR luminaire.

luminaire. A simpler design is to make a general module with an $M \times N$ array of white LED. In this case, to enlarge the capacity, M could be 6 and N could be 4. Thus, in total there are 24 vacancies on the board. If 24 pieces of LED are used, the streetlight can provide up to 120 W, which is sufficient to support most needs in an LED streetlight. In the case of 100 W, one of the choices is to put 20 pieces of white LED on the 24 vacancies of the MCPCB board. The other solution is to put 24 pieces of white LED, but the operation power can be reduced to 4.2 W. The choice between the reduction of white LED number and of driving power relates to cost and operation lifetime. Flexible design is an advantage of the ALA luminaire.

The ALA is a design with a lens array, but the TAR is for reflector arrays. In contrast to a lens, a reflector is not sufficiently flexible to form the desired light pattern. Therefore, the design of the TAR luminaire is more complicated. Each reflector in the reflector array could be the same, but the eccentric amount of each white LED corresponding to the reflector could be different, as shown in figure 7.6. The weight of a TAR Luminaire is similar to that in an ALA luminaire, but an additional flat cover is needed to protect the LEDs. Besides, the reflector array of the TAR design needs high-reflective coating.

A comparison will be made for the inducing glare to a pedestrian among these three types of LED streetlight, when equivalent luminance calculation is applied. The age effect for the eyes of people at the ages of 40 and 60 are included. The results demonstrate that the ALA could cause relatively smaller glare for most viewing conditions. In addition to low cost, high design flexibility and other benefits, the ALA luminaire for LED streetlighting is the most popular design.

The advanced technology for white LED streetlighting will be not limited by the lens design. Even though the peanut lens is suitable for most roadways, it is difficult to make the light pattern geometrically fit well on the roadway. Therefore, an advanced design is to use a microlens array plate (MLAP) to reach the goal [4]. An MLAP can be used to shape a collimating light to a specific pattern. However, an MLAP is not applicable to reduce the divergent angle of the incoming light. Thus the design with use of a lens array starts from using a TIR lens to collimate the light from a white LED. The divergent angle of the light passing through the TIR lens needs to be calculated, so that the width of the illumination pattern can fit the width of the roadway. Then an MLAP is used to spread the light pattern in one dimension, which is along the roadway. The whole structure is shown in figure 7.7. However, due to possible slanted angle of the LED luminaire, the ground illumination pattern could be a trapezoid rather than a rectangle. To solve this problem, the MLAP is suggested to be parallel to the surface of the roadway, as shown in figure 7.8. Through proper design of the light pattern and the poling distance, the roadway

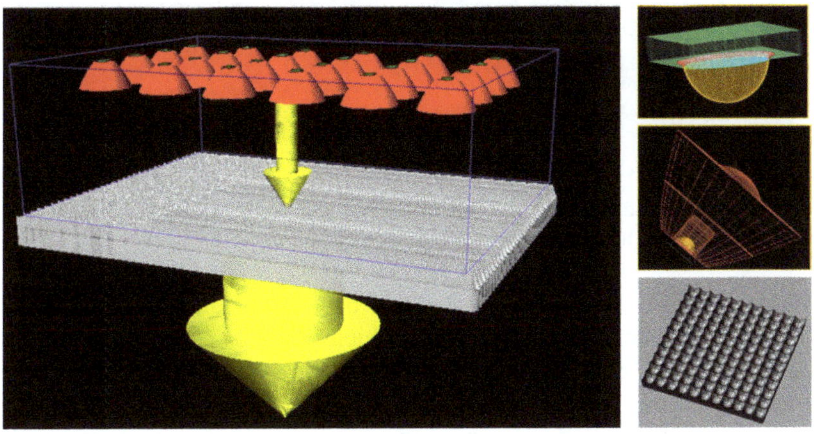

Figure 7.7. The structure of the LED streetlight with use of a microlens array plate. Reprinted with permission from [59].

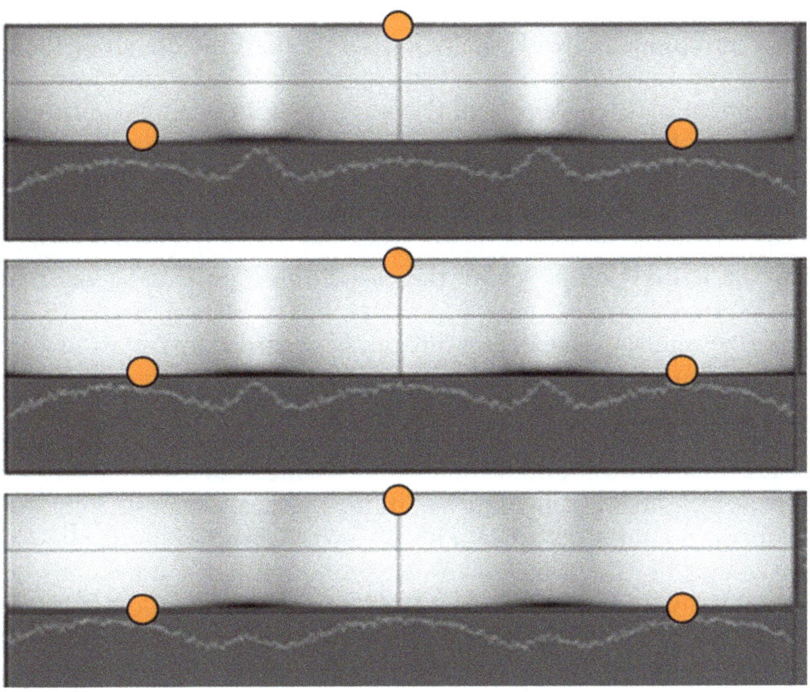

Figure 7.8. The illumination pattern on the roadway with different poling distances.

(a) (b) (c)

Figure 7.9. (a) A traditional streetlight, (b) a white LED streetlight, and (c) a white LED streetlight with use of a microlens array, where the luminaire looks dark.

illumination could be very uniform. Besides, the optical utilization factor of such a design could be as high as 60% or above, which is larger than 40% in the case with use of the peanut lens. Figure 7.9 shows photos of several streetlights, where the sky glow effect can be easily observed. Note that less sky glow means more efficient roadway illumination. The dark appearance of an LED luminaire is another guide to illustrate the utilization of LED light.

The advantage with the use of a microlens array is not only for spreading the illumination pattern in a flat rectangle shape. It is amazing to discover that the MLAP is applicable to shaping the illumination pattern for a curved roadway, which exists in mountain areas usually [5]. To do that, the design of the microlens array is the point. In the former case, each microlens has the same form. But in this case, the design of the microlens is more complicated. In order to uniformly connect the light pattern from one pole of streetlight to another, the illuminated curve pattern needs to be in a distribution similar to a Gaussian form. Therefore, a designer needs to design several kinds of microlenses as shown in figure 7.10, where the curve light patterns are with different sizes. The final combined illumination light pattern by the MLAP could spread the desired light pattern on the roadway. A schematic diagram shown in figure 7.11 shows that the design of an LED streetlight could adapt the roadway shape, and this can be regarded as the art of LED optics.

7.2 LED headlamps

White LEDs, owing to appropriate luminous existence and sufficient output flux, are suitable for vehicle forward lighting for train, truck, automobile, motorcycle, and bike. The forward light, also called a headlamp, is aimed to illuminate objects and the roadway ahead of the driver. Thus the projection flux must be high enough to provide good visibility to a driver. But the forward light should not cause glare to drivers or pedestrians on the roadway. Therefore, every country has its regulations for the vehicle headlamp to indicate the illumination in the bright zone and dark zone as well.

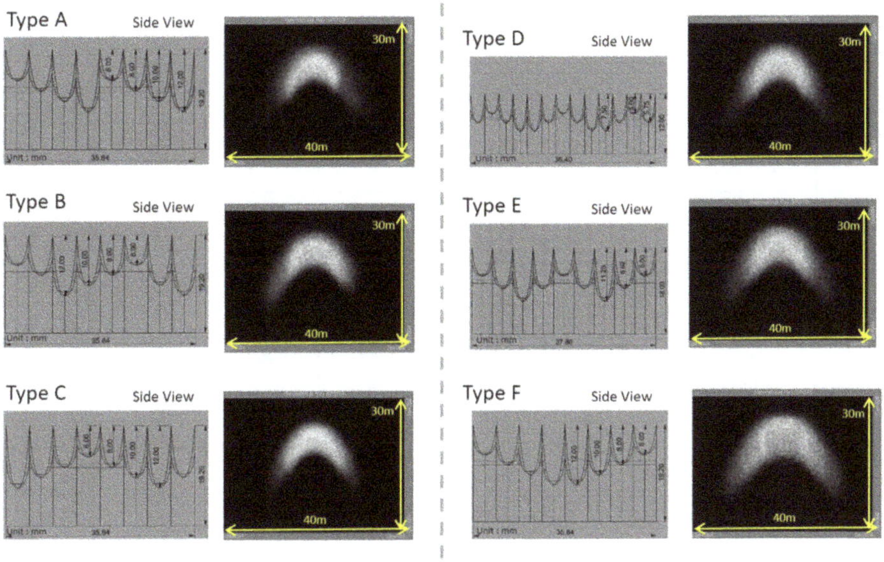

Figure 7.10. The geometry of the microlens and the corresponding light pattern on the ground. Reprinted with permission from [59].

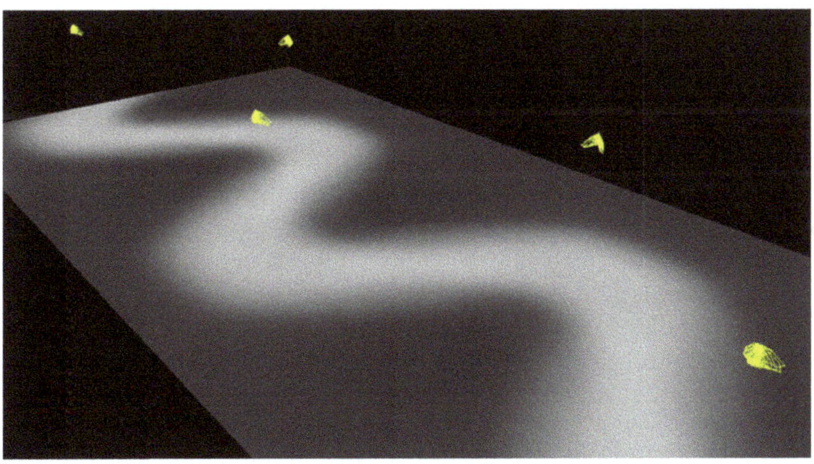

Figure 7.11. A schematic diagram of uniform illumination pattern with adaptive streetlight.

To design a white LED headlamp, a designer must be familiar with the associated regulation [6–9]. Essentially, headlamps can be classified as low beam, high beam, fog lamp, and daytime running light, as shown in table 7.1. Among those headlamps, the low beam is the most essential lamp, which is responsible for forward illumination for the driver, but a high-contrast cutoff line is required to avoid glare. A high beam is an auxiliary light, which provides bright illumination without the cutoff line, so the high beam should not be turned on when there is any vehicle or pedestrian ahead. A fog lamp is also an auxiliary lamp, which is helpful to a driver

Table 7.1. A list of the characteristics of various types of headlamp.

	ECE Regulation	Cut-off Line Requirement	Contrast Ratio	Hot Spot Requirement	Light Pattern at 25m
Daytime Running Light	ECE R87	NO		400 cd	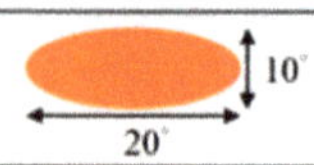
Fog Lamp	ECE R19	YES	1.5	1.5 lux	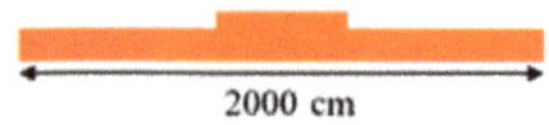
Motor-Low Beam	ECE R113	YES	4.3	3 lux	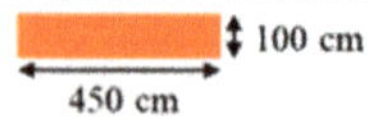
Motor-High Beam	ECE R113	NO		32 lux	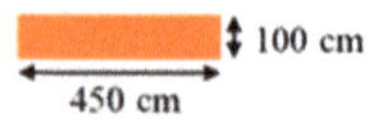
Auto-Low Beam	ECE R112	YES	30	16 lux	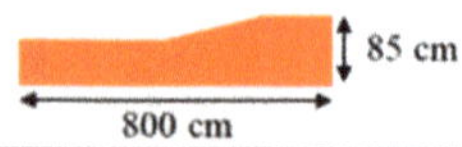
Auto-High Beam	ECE R112	NO		48 lux	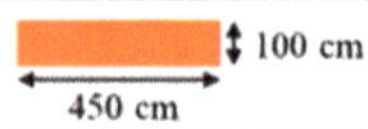

when the vehicle is in a fog circumstance. The fog lamp is equipped with a cutoff line and a longer horizontal illumination range is needed to illuminate those objects close to the vehicle. The daytime running light is to provide a signal rather than illumination for a vehicle. The objective of a daytime running light is to make the vehicle more visible to others. Therefore, there is no cutoff line, but the largest luminous intensity is limited.

Figures 7.12 and 7.13 show the detailed instructions of ECE R112 regulation for an automobile. In the part of the low beam, there are four checking zones. Zone I is the illumination region below the horizontal line. Zone IV is one of illumination regions in conjunction with Zone II. The points at 75R and 50R located at Zone II are the major bright areas to form the cutoff line. Zone II is the region limited by the horizontal line, Zone I, Zone IV, and the right/left vertical lines at 9°. There are several special requirements in the regulation to make the brightness correct at the setpoints. The high beam in ECE R112 regulation is a different story. As shown in figure 7.13, the high beam is aimed at long-range illumination and provides clearer and brighter vision. The light pattern is symmetrical with a central maximum.

In addition to automobiles and motorcycles, there are different vehicles, including bikes and e-bikes. The characteristics of a bike headlamp are light weight, compactness, low power consumption, and low cost because a bike is not equipped with a headlamp when it is sold to a user. An e-bike is different from a traditional bike. There is a big cell to provide electricity, so the e-bike is equipped with a headlamp. The German regulation of a bike (called K-mark regulation) is shown in figure 7.14, and figure 7.15 is for an e-bike under ECE Class B regulation that requests higher sharpness and linearity (here called E-mark regulation). In the regulations, the low beam requirement of the minimum illuminance at A point is 20 lux for the K-mark

(a)

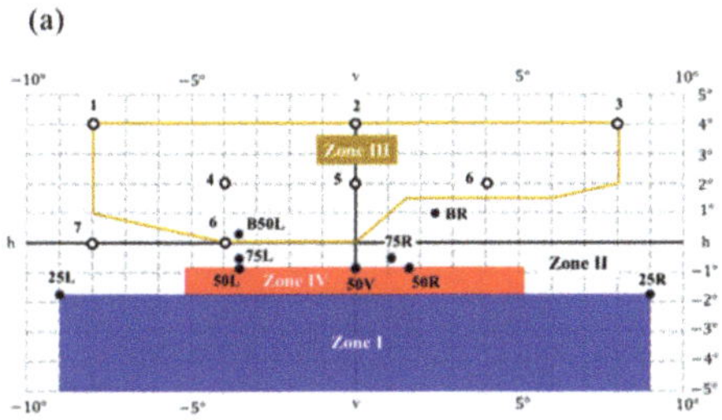

(b)

Test point	Angular Coordinates [Degrees]							Required luminous intensity [cd]
B 50 L	0.57U, 3.43L							< 350
BR	1.0U, 2.5R							< 1750
75 R	0.57D, 1.15R							> 10100
75 L	0.57D, 3.43L							< 10600
50 L	0.86D, 3.43L							< 13200
50 R	0.86D, 1.72R							> 10100
50 V	0.86D, 0							> 5100
25 L	1.72D, 9.0L							> 1700
25 R	1.72D, 9.0R							> 1700
Any point in zone III	Bounded by the following coordinates							< 625
	8L	8L	8R	8R	6R	1.5R	V-V	
	1U	4U	4U	2U	1.5U	1.5U	H-H	
Any point in zone IV	086D to 1.72D, 5.15L to 5.15R							> 2500
Any point in zone I	1.72D to 4D, 9L to 9R							$< 2 \times I_{50L}$ or $2 \times I_{50R}$
1 + 2 + 3	(4U,8L) + (4U,0) + (4U,8R)							> 190
4 + 5 + 6	(2U,4L) + (2U,0) + (2U,4R)							> 375
7	0, 8L							> 65
8	0, 4L							> 165

Figure 7.12. (a) The important tested zone of the low beam of ECE R112 regulation, and (b) the detailed definition.

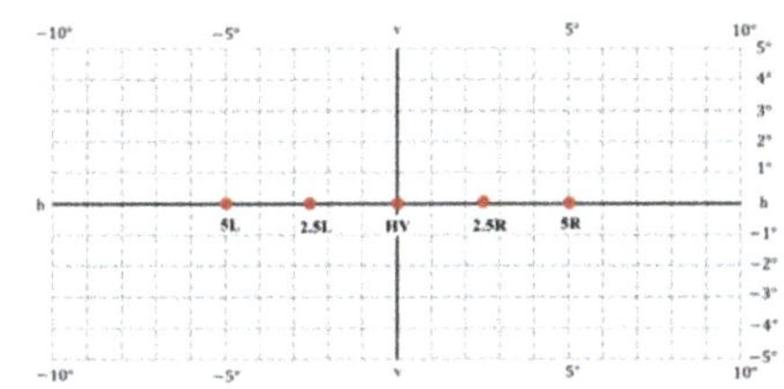

Test point	Angular Coordinates [Degrees]	Required luminous intensity [cd]
I_{max}		$215{,}000 > I_{max} > 40{,}500$
H-5L	0.0, 5.0L	>5,100
H-2.5L	0.0, 2.5L	>20,300
H-2.5R	0.0, 2.5R	>20,300
H-5R	0.0, 5.0R	>5,100
$HV \geq I_{max} \times 80\%$		

(a) (b)

Figure 7.13. (a) The important tested zone of the high beam of ECE R112 regulation, and (b) the detailed definition.

regulation and at 50 V is 22 lux for the E-mark regulation. The high beam requirement for the HV point is 50 lux for the K-mark regulation, and 160 lux for the E-mark regulation. In addition, the E-mark regulation requests more than 200 lux at the brightest point in the high beam mode. Such a condition makes the design in the E-mark regulation more complicated than that in the K-mark regulation.

There are several approaches in optical design for a headlamp, including using a reflector, a multi-segment lens, and the hybrid structure. For an automotive low beam, the illumination must be stronger to fit the regulation. A typical approach is to use a hybrid optical element, which contains an imaging lens and an elliptical reflector [10, 11]. As shown in figure 7.16, the white LED is located at one of the focal points of the elliptical reflector. The reflected light will be focused around the other focal point, where a baffle with a special shape is located. The focused light passing the baffle will be shaped into a special form. An imaging lens is used to project passing light away and thus the illumination pattern can fit the regulation at

(a)

Position	Angular coordinates (degrees)		Illuminance (lux)
A	3.5°D	V	> 20 lx
From CR to CL	3.5°D	4°L ~ 4°R	≥ E_A/2
From A to B(included)	3.5°D ~ 5°D	V	>10 lx
From B to M	5°D ~ 8.5°D	V	> 3 lx
From GL to GR	8.5°D	8°L ~ 8°R	> 2 lx
HV	On the H-H line and above H-H line		≤ 2 lx
Zone 1 illumination values shall be ≤ 1.2 E_A Zone 2 illumination values shall be ≤ E_A			

(b)

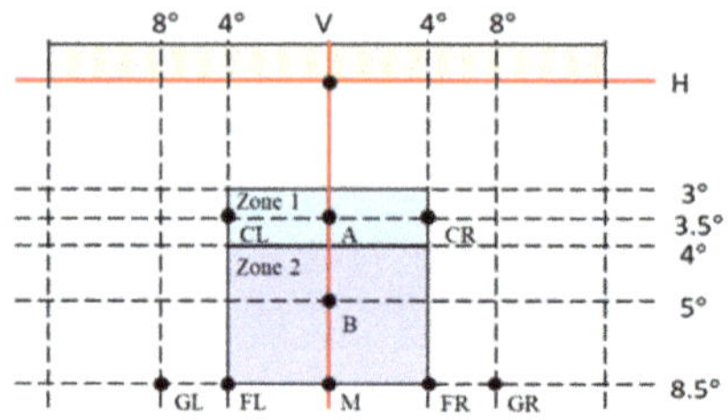

(c)

Position	Angular coordinates (degrees)		Illumination values in lux
HV	H	V	≥ 50 lx
A1 and A2	H	4°L and 4°R	≥ E_{HV}/2
A3	3.5°D	V	≥10 lx

(d)

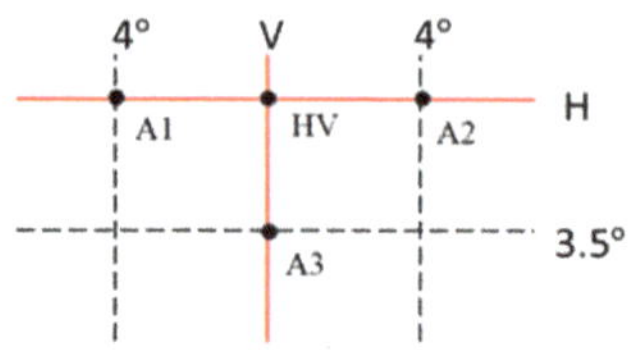

Figure 7.14. (a) The important tested zone of the low beam of K-mark regulation, (b) the detailed definition, (c) the important tested zone of the high beam of K-mark regulation, and (d) the detailed definition.

(a)

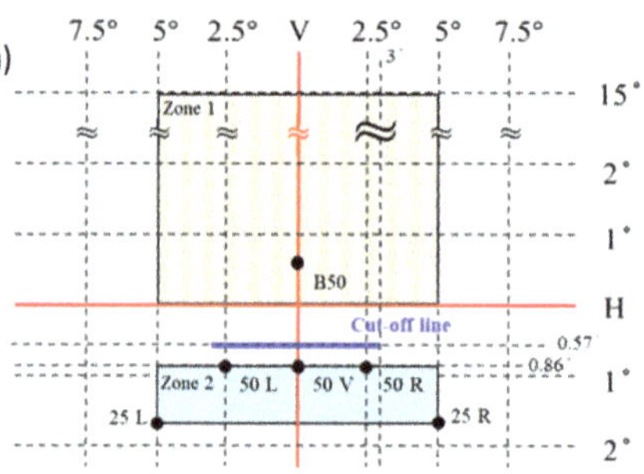

(b)

	Angular coordinates (degrees)		Luminous intensity (cd)
Any point in zone 1	0°~15°U	5°L ~ 5°R	≤ 700 cd
Any point on line 50L to 50R	0.86°D	2.5°L ~ 2.5°R	≥ 1,100 cd
Point 50 V	0.86°D	V	≥ 2,200 cd
Any point on line 25L to 25R	1.72°D	5°L ~ 5°R	≥ 2,200 cd
Any point in zone 2	0.86°~1.72°D	5°L ~ 5°R	≥ 1,100 cd

(c)

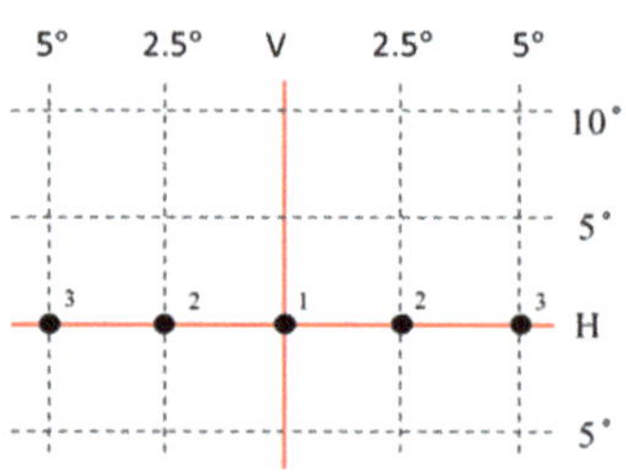

(d)

Test point number	Angular coordinates		Luminous intensity
1	H	V	≥ 16,000 cd
2	H	2.5°R and 2.5°L	≥ 9,000 cd
3	H	5°R and 5°L	≥ 2,500 cd
Minimum of $I_{maximum}$ = 20,000 cd Maximum of $I_{maximum}$ = 215,000 cd			

Figure 7.15. (a) The important tested zone of the low beam of E-mark regulation, (b) the detailed definition, (c) the important tested zone of the high beam of E-mark regulation, and (d) the detailed definition.

the defined distance. Figure 7.16(b) shows an advanced design, where a beam shaper is used to replace the baffle. The advantage is that the blocked light in the case shown in figure 7.16(a) can be reused to enhance the brightness in the bright zone. Figure 7.17 shows an alternative design, which can multiplex multiple LEDs with

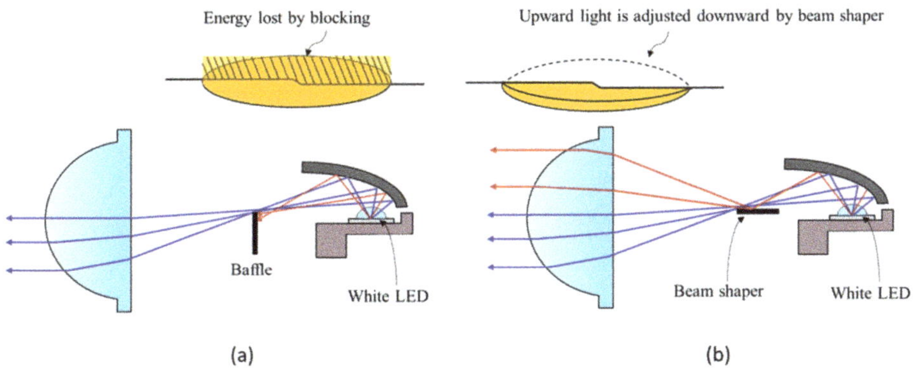

Figure 7.16. (a) The traditional design with an elliptical mirror and a baffle to form the cutoff line, and (b) the baffle is replaced by a beam shaper [10, 11].

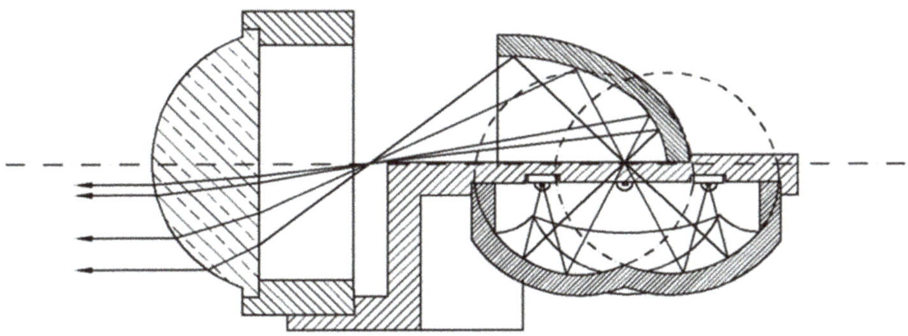

Figure 7.17. A design using merged elliptical mirrors [12, 13].

merging elliptical reflectors [12, 13]. Figure 7.18 shows a design with a solid freeform optical element [14]. The LED is focused at a location through total internal reflection by the elliptical boundary. After beam shaping by the special shape at the location, the light is projected by the front lens. Such a design could face fabrication difficulty when the white LED is immersed in the optical element. Theoretically, there are simpler ways to form the regulation pattern. Figure 7.19 shows that a baffle is located at the front end of the LED array [15]. In comparison with those designs with a baffle or a beam shaper, the design shown in figure 7.16(b) could be more energy efficient and practical.

Another approach is to use a light pipe to form the light pattern before sending it to the projection lens, as shown in figure 7.20 [16–18]. To avoid serious optical aberration, a similar design shown in figure 7.21 is to tilt the outer ring of the projection lens to direct the light on the outer part on the lens downward for ground illumination.

The third approach is to use a multi-segment reflector to form the regulation pattern [19]. Using a multi-segment reflector is one of the advantages of white LEDs because the light source is relatively small and always faces one side. Thus the white

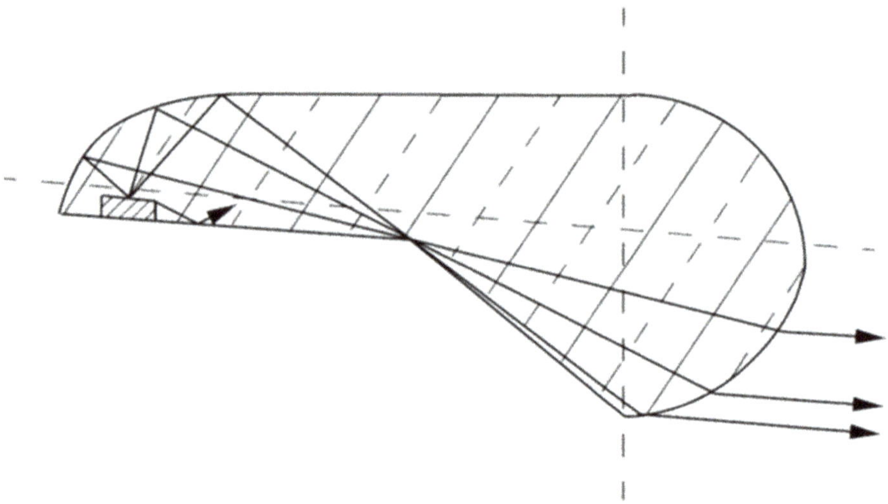

Figure 7.18. A design using volume type of freeform lens [14].

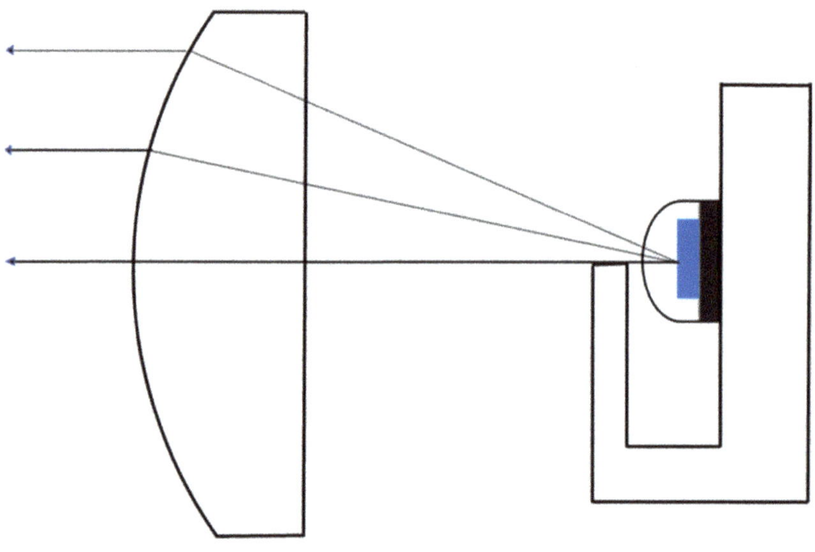

Figure 7.19. A baffle is located at the front end of the LED array to form the cutoff line [15].

LED could be placed at the upper side or bottom side of a reflector, as shown in figure 7.22. Through a precise mid-field optical model of the light source and an appropriate illumination design [20–22], the curvature and orientation of each segment can be determined by following the procedure suggested in table 7.2. The merged light pattern can fit the regulation. Figure 7.23 shows a designed structure of a multi-segment reflector for ECE Class B regulation and the simulated light pattern. Through proper machining and fabrication, the real light pattern by the

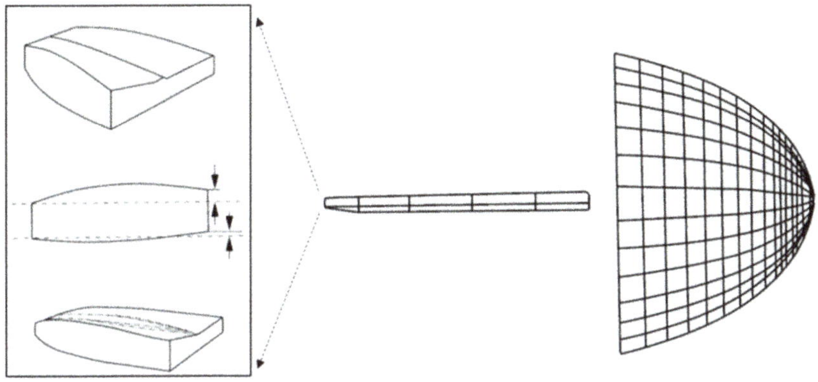

Figure 7.20. A light guide with an imaging lens is used to form the designed cutoff line [16–18].

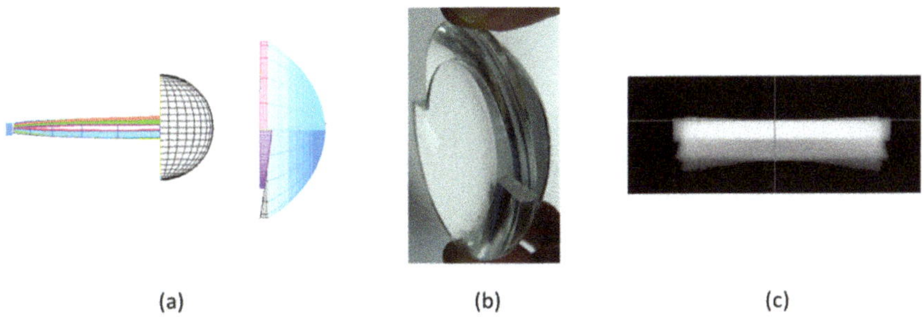

Figure 7.21. (a) A design similar to that in figure 7.20, but the lens is modified, (b) a prototype of the lens, and (c) the light pattern. Reprinted with permission from [60].

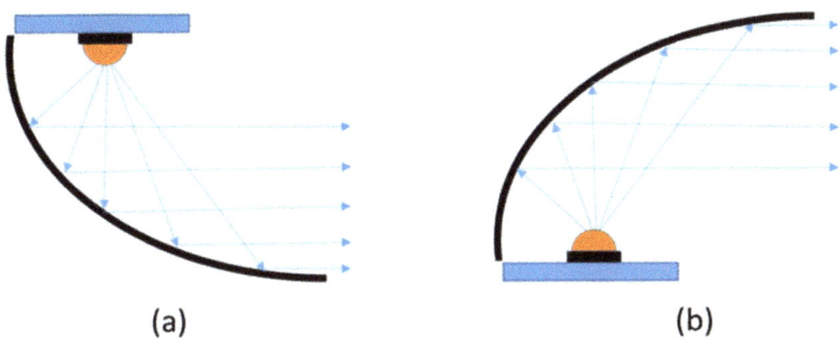

Figure 7.22. (a) A design with LED downward lighting, and (b) upward lighting [19].

prototype could fit well the design and meet the requirement in the corresponding regulation.

For a bike headlamp, the design concept is different from an automotive headlamp. The headlamp should be compact, lightweight, with low power consumption, and low cost. Figure 7.23 has shown an example of an e-bike headlamp.

Table 7.2. The flowchart for designing an LED headlamp.

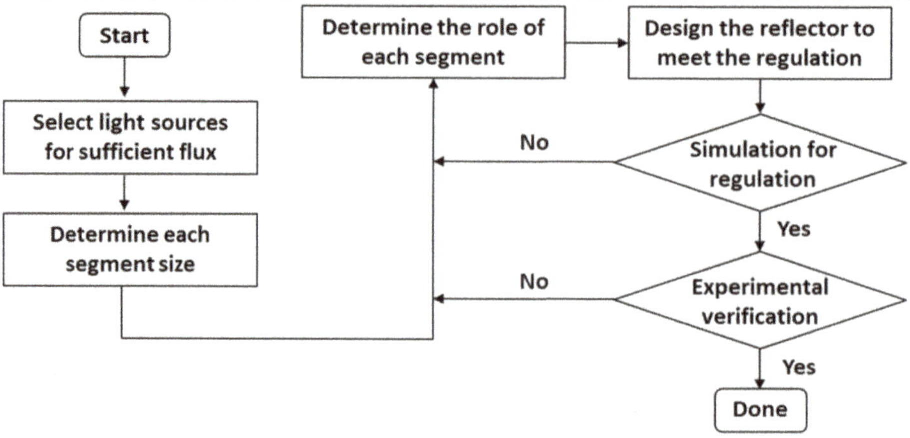

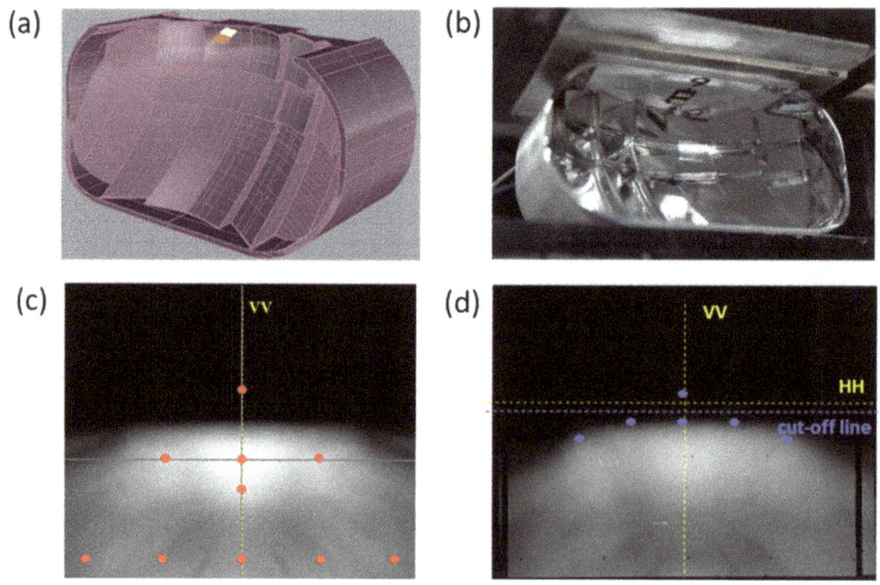

Figure 7.23. (a) A designed structure of a multi-segment reflector for ECE Class B regulation, (b) the prototype, (c) the simulated light pattern, and (d) a photo of the light pattern at 10 m.

A multi-segment reflector can project a satisfying low beam, and even a high beam at the same time, as shown in figure 7.24. Besides, roadway illumination is provided. Such a design is complicated, and the role of each segment must be well defined. In most conditions of a conventional bike head lamp, meeting K-mark regulation represents high quality, where the requested illumination at 10 m for the bright point

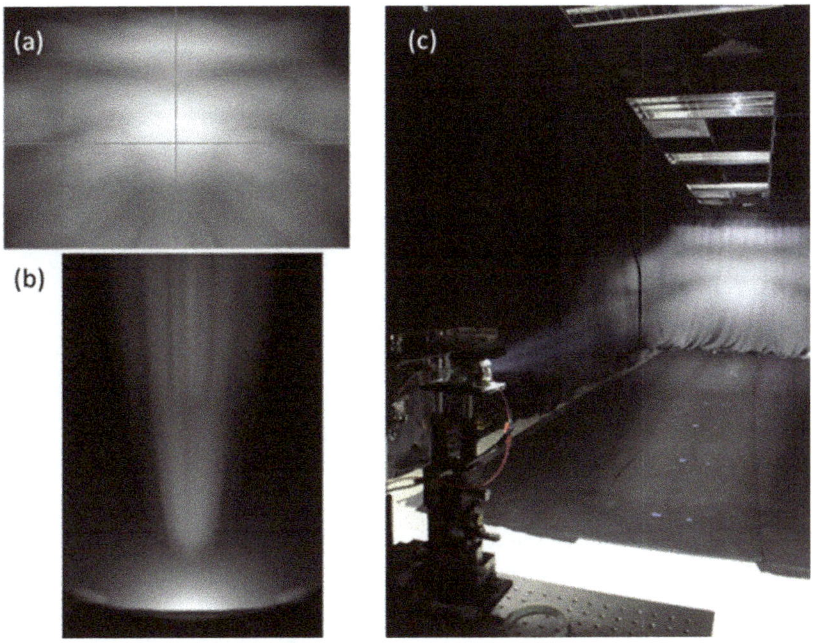

Figure 7.24. (a) The simulated low-beam and high-beam pattern, (b) the ground illumination, and (c) the total field observed in the experiment. Reprinted with permission from [61].

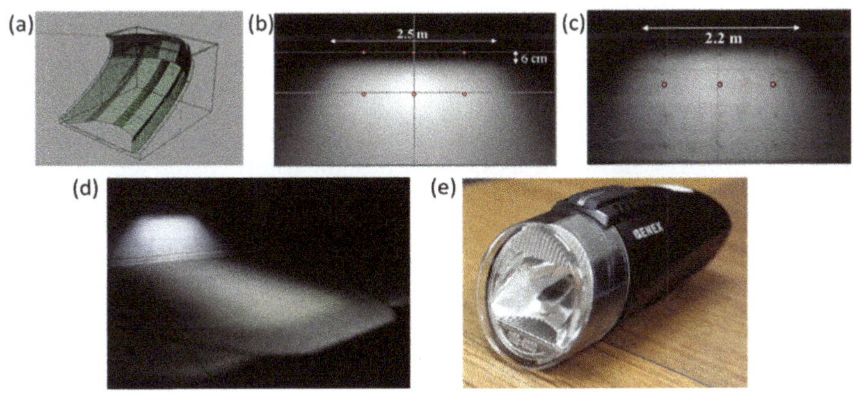

Figure 7.25. (a) A design for bike headlamp, (b) the simulated light pattern, (c) the photo of the light pattern, (d) the complete light pattern in the experiment, and (e) the looking of the bike headlamp.

(A point) is only 25 lux. Thus using a white LED operated 1 W to 3 W is enough to reach the objective. Figure 7.25 shows a successful design of a bike headlamp with low power consumption, where the driving power on the white LED is only up to 2.5 W [23]. The design is also simple, where there are eighty segments in the reflector. The upper four segments are used for ground illumination, and the four lower segments are used to form the high-contrast cutoff line. The result is that not only is

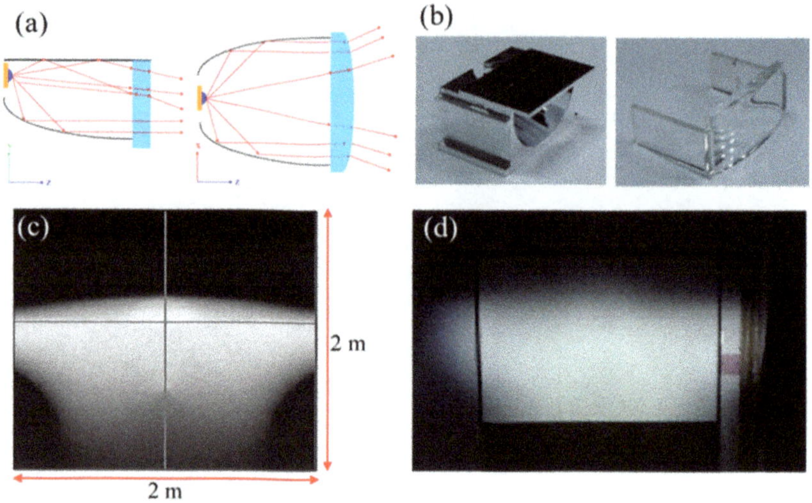

Figure 7.26. (a) The geometry of the truncated reflector with a cylindrical lens. (b) the prototype, (c) the simulated light pattern, and (d) the experimental observation.

the K-mark regulation met but also very good illumination on the roadway from 1 m to 10 m ahead of the driver is well performed.

Since low cost is one of the key issues in a bike headlamp, simple but effective design is always demanded. Figure 7.26 shows a simple design to meet K-mark regulation [24]. The design starts from using a truncated reflector to collimate the LED light and using a cylindrical lens to laterally spread the light pattern. In order to have a high-contrast cutoff line, the reflector is truncated on the upper part. The prototype projects a light pattern similar to simulation at 10 m and the light pattern meets the K-mark regulation. Using a light pipe is another strategy to meet the regulation. Figure 7.27 shows an associated design, where a step-shape light pipe is applied to form the illumination pattern [25]. R1 in the light pipe directs the incoming light to R2, which is used to form a laterally spreading light pattern. Meanwhile, R1 is also a shading plate to block the upward light to the dark zone in the regulation, and is helpful to form the high-contrast cutoff line. R3, near the light source, is used to redirect the incoming light downward to the ground.

The design of an automotive is more feasible without those specific constraints in a bike. Therefore, a headlamp set could combine several different optical elements to perform the best illumination proposed by the carmaker. Figure 7.28 shows an example of an automotive headlamp set, where a simple reflector, several multi-segment reflectors, and reflector–lens combination elements are put together to perform low beam and high beam requested by the regulation. Owing to the advanced development in LED technology, a matrix type of light source can be made. The matrix type of LED uses tiny LEDs (e.g., mini LED) to build up a spatially controllable or dimmable light source. Such a matrix-type LED module can be regarded as a spatial light modulator, which is the panel for an optical projector. If the headlamp has a function similar to a projector, the projected light pattern can be

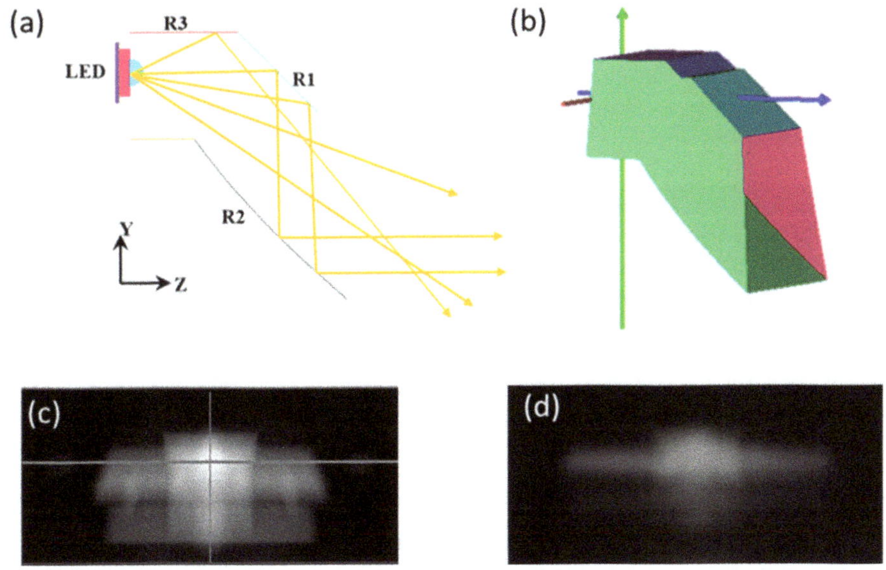

Figure 7.27. (a) The geometry of the step shape light pipe, (b) the 3D model, (c) the simulated light pattern, and (d) the experimental observation.

Figure 7.28. A prototype of automotive headlamp.

adjusted. Several carmakers utilize this characteristic to develop adaptive headlamps [26–32], where the projection light pattern can be automatically adjusted upon the roadway condition sensing. The headlamp function shown in figure 7.29 demonstrated by BMW provides a big benefit of the largest visibility to the car driver [33]. The adaptive headlamp is turned on at the high beam mode with local dimming to prevent glare to other drivers on the same roadway. While the other drivers are not influenced by the adaptive headlamp, the driver in the car has the biggest illumination

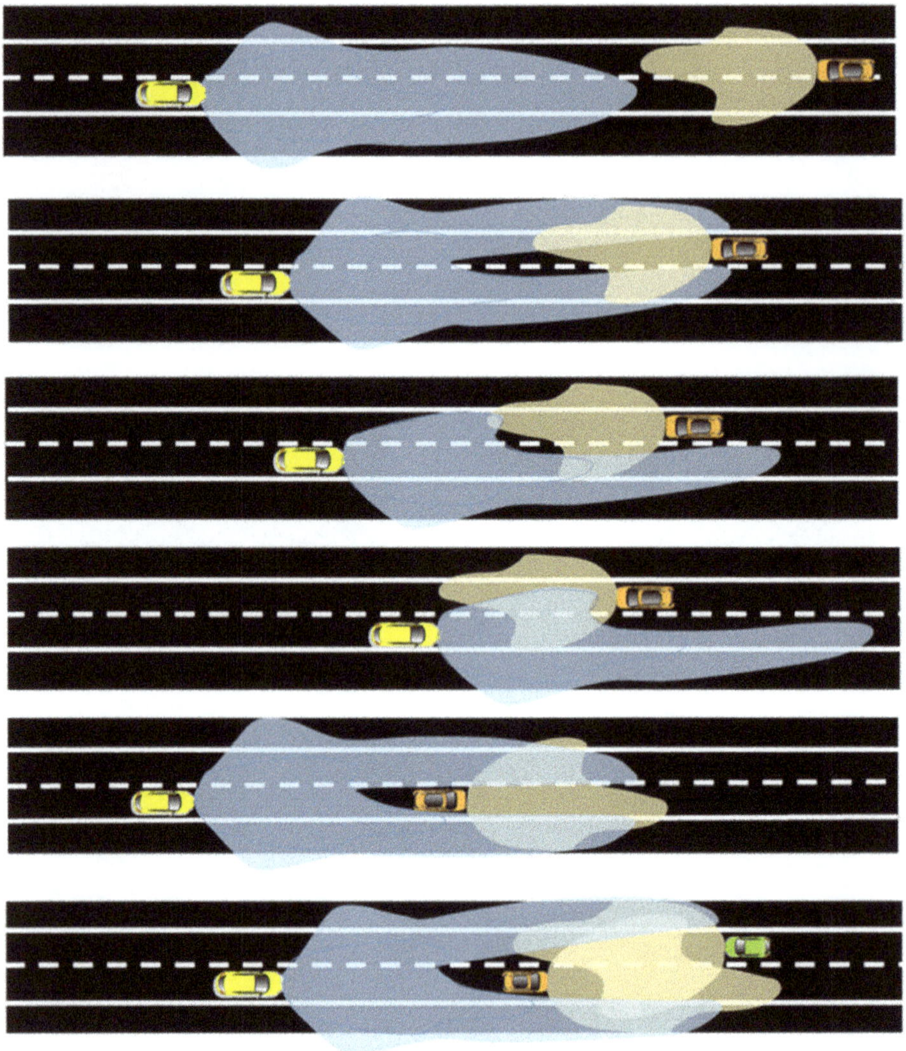

Figure 7.29. The concept of BMW's adaptive headlamp [33].

on the roadway at the same time. Figure 7.30 shows an approach with use of an LED matrix to serve as a spatial light modulator, which can be regarded as a light modulator to encode a 2D pattern and is projected by an imaging lens [34]. The illumination pattern is in the conjugate plane of the LED matrix, so the dimming part of the LED will correspond to the adaptive light pattern in the illumination.

The adaptive headlamp can also perform by laser illumination through fast scanning and lens projection. The laser light is at a wavelength similar to the blue LED in solid-state lighting. When a blue laser light passes through a phosphor plate, some of the blue light will be absorbed by the phosphor, and yellow light will be emitted [35–42]. Finally, the illumination of the blue laser will behave as a new light

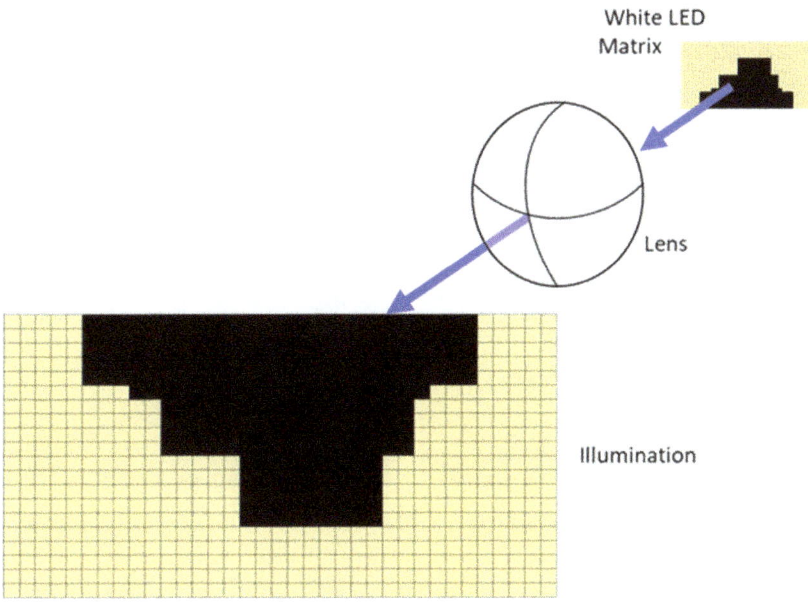

Figure 7.30. An LED matrix is imaged to the illumination plane with local dimming function.

Figure 7.31. A blue laser illuminates a yellow phosphor plate.

source to emit white light, as shown in figure 7.31. Figure 7.32 shows experimental results with laser scanning for automotive forward lighting. In the experiment, there are four simulated pedestrians facing the headlamp, where there is a camera to detect an object such as a pedestrian or car in front of the vehicle. Digital information processing or other advanced technology (for example, artificial intelligence (AI) technology) could recognize the object and decide to turn off a

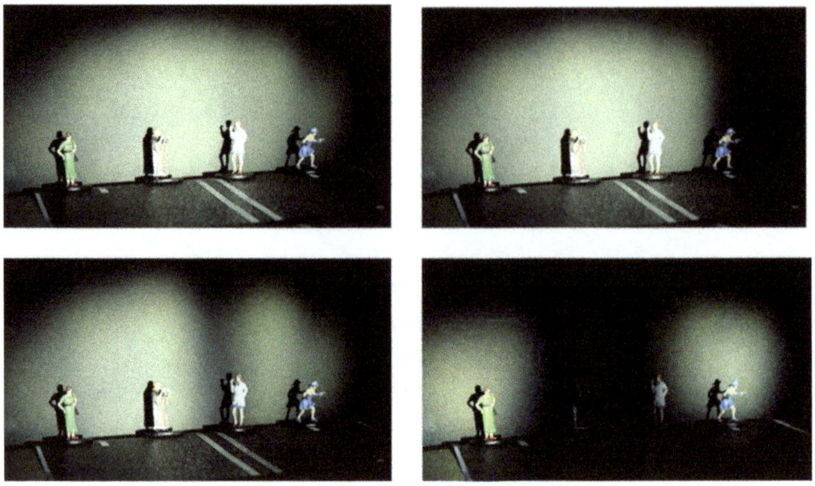

Figure 7.32. A demo of adaptive illumination with laser scanning white light. Reprinted with permission from [62].

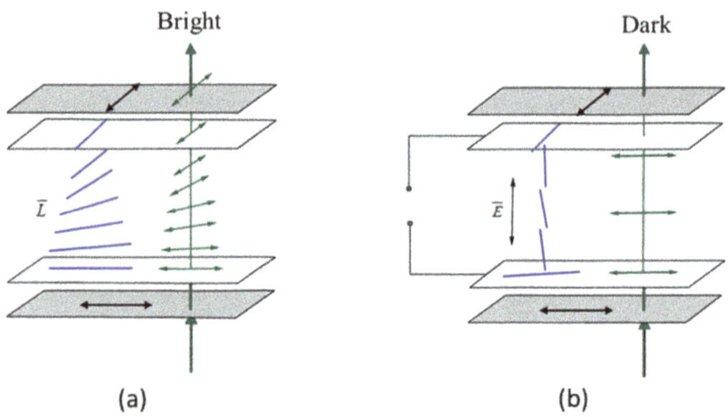

Figure 7.33. (a) Evolution of polarization for an LCD without applied electrical field, and (b) with an applied electrical field.

certain part of the illumination pattern. Such an adaptive headlamp system will play a more and more important role in automotive forward lighting.

7.3 LED backlight

Liquid crystal display (called LCD) is the most popular flat display, which could be a large display panel with a size as large as 85 inches or above,w or a size of 1 inch containing two million pixels. Applied to display, a liquid crystal (LC) panel is a device which is used to control the polarization state of the incoming light and finally control the transmission efficiency, as shown in figure 7.33. However, the liquid crystal panel is a passive component, which does not emit light. To provide sufficient brightness of an LCD, a backlight system is needed. The backlight system is aimed to provide uniform and bright illumination to an LC panel. The structure can be

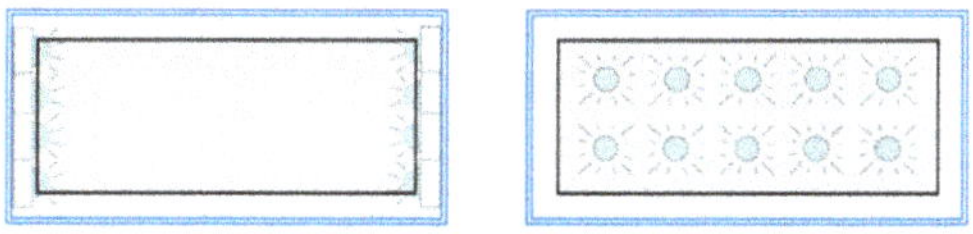

	Side-view	Direct-view
Efficiency	70%~80%	>70 %
Optical uniformity	Good	Good
Heat dissipation	Fair	Good
Optical design (BL)	Hard	Easy
Angle required (LED)	Narrow angle	Wide angle
Panel size	Small	Large
Thickness	Slim	Thicker
Key component	Light guide	Diffuser

Figure 7.34. A comparison between the side-view and the direct-view backlight structures.

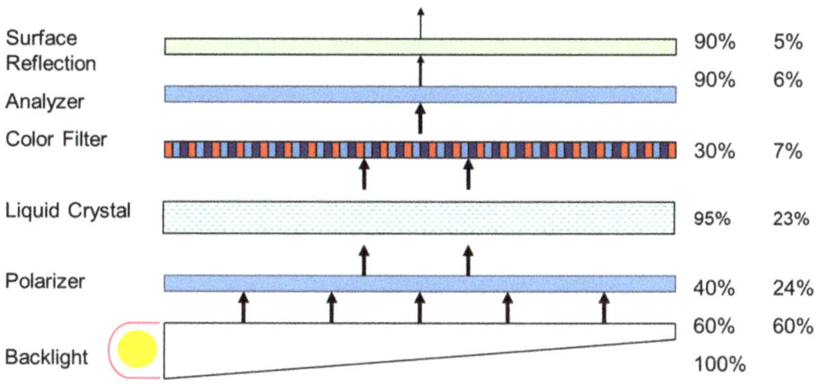

Figure 7.35. A structure of side-view backlight with a light guide plate.

divided into two categories. One is a side view and the other is direct view, as shown in figure 7.34 [43].

Side view is a way to incorporate a light source and a light guide plate to provide a 2D uniform light source. As shown in figure 7.34, the light source is required to have a narrow beam angle to couple into a light guide. The design of the light guide is complicated. There are several kinds of designs for the light guide plate. One is to make the light guide plate with a wedge, and then the coupling light from one side can be effectively re-directed to the appointed direction, as shown in figure 7.35. Another is to encode tiny print spots across the back side of the backlight plate. The distributed spots are used to scatter the incident light to the front exit face. Design of the distributed spots must consider the light field by the light source. In the past, when the LC panel was applied to a monitor or a TV, the light source for the backlight was the cold cathode fluorescent lamp (CCFL), which contains mercury and phosphors. The high voltage electrons collide with the mercury and then UV

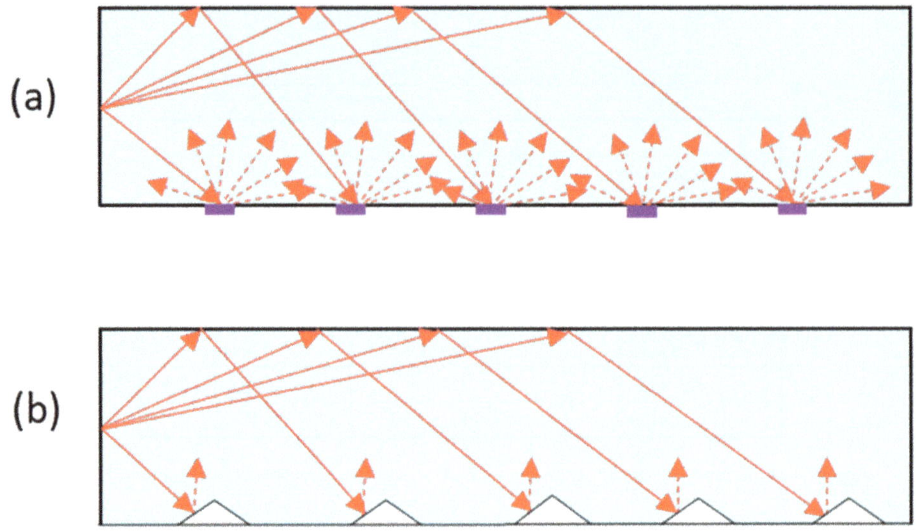

Figure 7.36. (a) The ray paths for a printed backlight patterning, and (b) for a v-cut structure.

light is emitted to excite phosphors. Through a proper design, the CCFL can be regarded as a linear light source, which is placed one side of the backlight plate. To redirect the incoming light toward the normal of the panel face as shown in figure 7.36, a structure called v-cut is used to replace the tiny print spot. Owing to the advantages of LEDs, the CCFL is replaced by white LEDs when the LCD panel is applied to a mobile phone, where the size is always smaller than 6.5 inches. In addition to the size of the light source, the CCFL is no longer competitive because it contains mercury and is relatively fragile.

In comparison with a CCFL, white LEDs are more flexible in geometry, lower power consumption, and higher color saturation. Since all rays are incident on one side of the light guide plate, the white LEDs used for the light source should be arranged in a linear array. The design of light coupling of white LEDs is different from that of a CCFL. The major difference is the discrete distribution of the LEDs. The light discontinuity forms a triangle dark zone, as shown in figure 7.37. In order to remove or squeeze the dark zone length, one approach is to form a coupling structure like a cylindrical lens array at the side entrance face to spread the coupling light laterally. The other approach is to put more LEDs on the side entrance face and to reduce the spacing of the LEDs. Generally, the side-view approach is an effective way to reduce the thickness of the LCD and is also cost effective.

The second approach for the backlight kit is direct view, where the light sources are placed behind an LC panel, as shown in figure 7.38. To make sure the brightness across the exit face of the LCD is uniform, the light sources must be put in a two-dimensional arrangement. Originally, the CCFL was the best light source for this purpose. A linear array of CCFL will construct a 2D light source. But the spacing between each two CCFLs is still a problem. To ensure uniform brightness,

Figure 7.37. The dark zone appears in the entrance region.

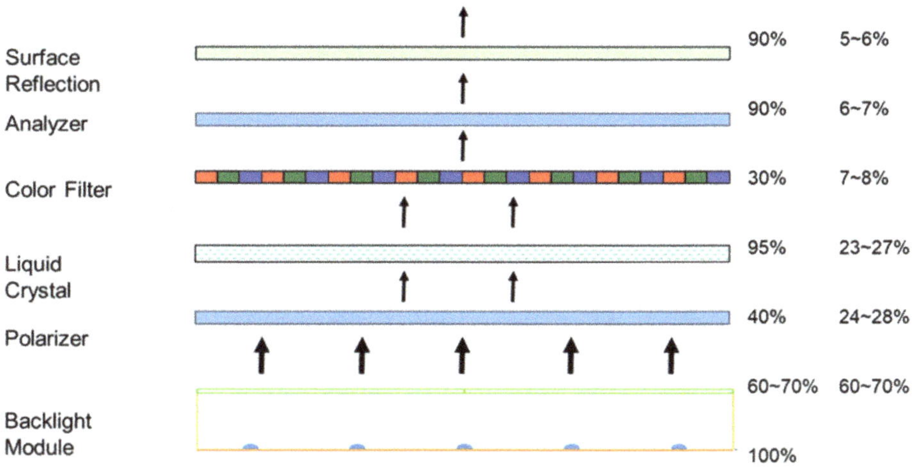

Figure 7.38. A structure of direct-view backlight with LED array.

increasing the thickness of the backlight system or using more CCFLs to reduce the spacing is an effective way. But the CCFL is no longer suitable for LCD because the CCFL contains mercury and is not very functional in the advanced backlight technology. White LEDs again replace CCFLs in the backlight system in the direct-view approach. The reason is not in cost, but in technology. When an LCD system contains a direct-view backlight kit, it means that they could embed advanced technology. First, if a white LED array with small spacing is installed for the backlight kit, not only can the thickness be reduced but also the LCD could have a narrower side bar around the LCD. Besides, the high-density LED array could be regarded as a secondary display behind the LC panel. The secondary display is not for display purpose but for enhancement of the display quality, including high dynamic range of brightness and motion blur reduction. In comparison to the OLED display, light leakage in a dark zone is an inherent shortcoming of an LC panel. Generally, it is difficult for an LCD panel to block all the light passing through the panel. This property limits the dynamic range of brightness.

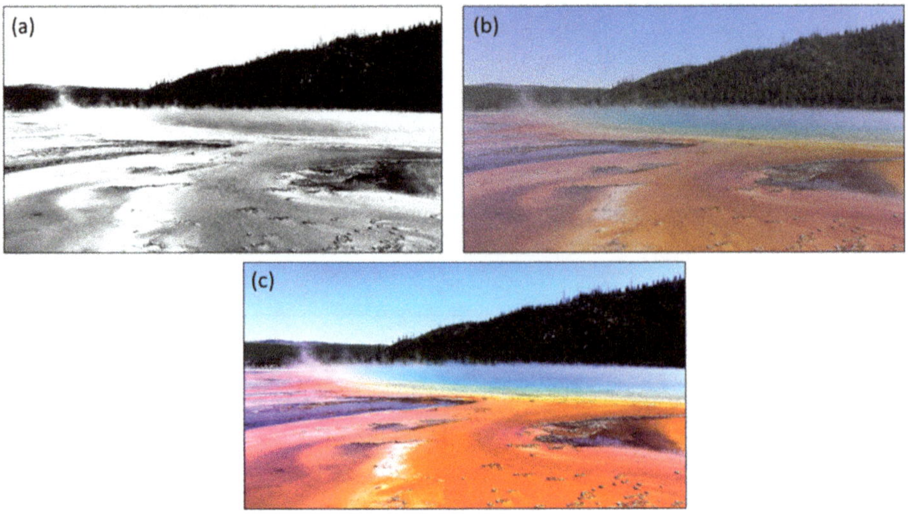

Figure 7.39. (a) The backlight plate serving as a secondary display panel, (b) a display without local dimming, and (c) a display with the backlight shown in (a) to perform local dimming.

To increase the contrast of the display and reduce the light leakage in the dark zone, a different approach for direct-view backlight offers the solution. As shown in figure 7.39, the secondary display by the LED backlight kit can perform local dimming for an LC panel. When there is a dark zone in the display, the corresponding area in the secondary display will be turned off to stop light from penetrating through the LC panel. Thus, the brightness in the dark zone of the display will be further reduced. Research reports show that local dimming technology supported by the secondary display of the LED backlight kit could increase the contrast from 10 000 to 1 000 000 or above. If the white LED is made by RGB LED x-in-one package, local dimming can be upgraded to color purity enhancement. Such an approach could provide a higher color gamut. The second benefit by the direct view approach is that the frequency of the sequential lighting across the LED array can fit the response speed of the LC panel to reduce motion blur owing to the slow response of LC molecules.

To design an LED backlight kit, the power of brightness management is important [43]. The LCD is a polarization sensitive device. A liquid crystal device can rotate the polarization direction of the incident light through an applied electrical field and a polarizer is used to transfer the polarization state to brightness. However, most of the light sources are not equipped with polarization components so that the emitted light is unpolarized. Therefore, there is a linear polarizer in front of the LC panel to make the incident light linearly polarized. Such an arrangement is cost effective and useful but wastes half the energy of the incident light. To solve this problem, 3M provides a functional film called dual brightness enhancement film (DBEF). The DBEF contains multiple anisotropic layers. For incidence of unpolarized light, the DBEF will reflect the light with a certain linear polarization (say P_2), and transmit the light with the other polarization (say P_1), as shown in

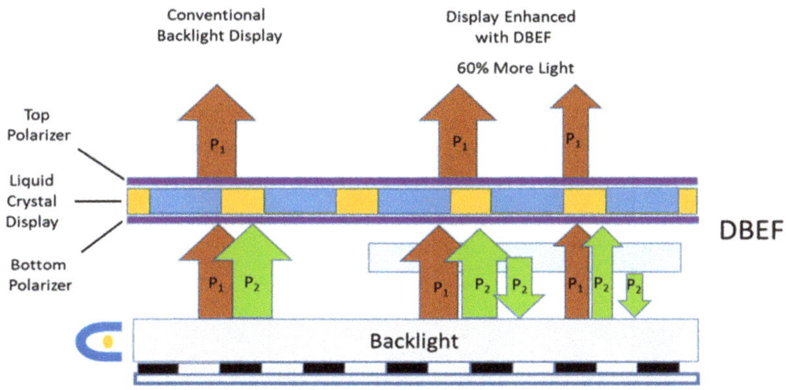

Figure 7.40. The concept of 3M DBEF for photon recycling.

figure 7.40. If the cavity of the backlight kit can depolarize the reflected light in the cavity, the recycling incident light could be extracted by the DBEF again, and the polarization state of the transmitted light will always be in P_1. Such an anisotropic film can not only serve the role of linear polarizer but also perform photon recycling to save energy.

The optical efficiency of the backlight kit is an important factor to the power management of an LCD. The power consumption of the LEDs to reach the desired brightness of an LCD can be estimated through calculating the luminous efficacy of LEDs, backlight cavity efficiency, transmission light distribution of the LCD, and the brightness gain by the multiple optical films. The total flux provided by an LCD can be expressed as [43]

$$\Phi = L \times A \times \frac{\pi}{G} = P \times \eta_{ecy} \times \eta_{BL} \times \eta_{TV}, \tag{7.1}$$

where L is the luminance of the LCD along the normal direction, A is the effective area of the LC panel, G is the total gain by the BEF (brightness enhancement film) and DBEF in comparison to a Lambertian light source, P is the injection electrical power of the LEDs, η_{ecy} is the luminous efficacy of each LED, η_{BL} is the optical efficiency of the backlight kit, and η_{TV} is the transmission efficiency of the LC panel. In a direct-view backlight kit, photon recycling will enhance the cavity transmittance, which is written [44]

$$\eta_{\mathrm{BL}} = T + TR_bR + TR_b{}^2R^2 + \cdots = \frac{T}{1 - R_bR}, \tag{7.2}$$

where T is the one-shot transmittance (OST) and R is reflectance of the diffuser, R_b is the reflectance of the surfaces in the cavity. Using equation (7.2), if the OST of the diffuser is around 70%, the reflectance of the diffuser is assumed as 28% and cavity wall reflectance is assumed as 90%, the cavity efficiency is as high as 93.6%. In comparison with 75% of efficiency of a side-view backlight, the calculated efficiency has a gain around 125%, so that such a design is more energy saving. If a diffuser

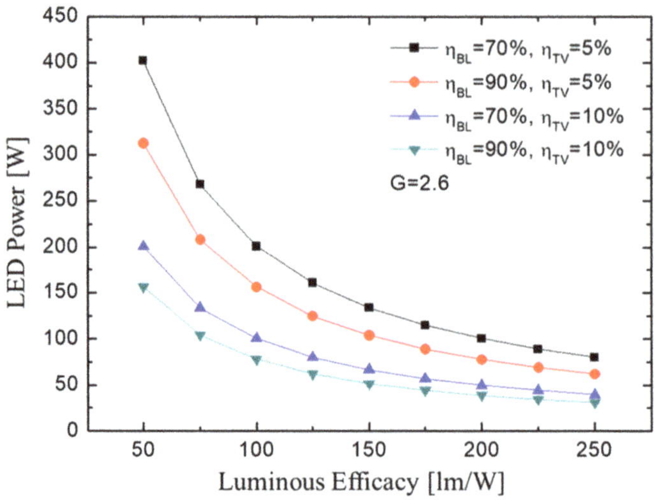

Figure 7.41. LED power versus luminous efficacy of a white LED under different backlight efficiencies and optical efficiencies of the display panel.

with a higher OST is used, the larger OST means more transparency, and higher cavity thickness is needed to hold the same brightness uniformity across the panel.

Energy saving is always an issue. The direct-view backlight with local dimming technique is one approach to save the energy in advance. But the most important way is to increase the luminous efficacy of white LEDs. In general, the typical transmission efficiency of an LC panel is between 5% and 10% [45–47]. However, a BEF can be used to increase luminance in the normal direction, and DBEF is another way of applying polarization recycling to increase optical transmission efficiency of a backlight kit. A calculation based on equation (7.1) for power consumption of white LEDs used for a 65-inch LCD at 500 nits is shown in figure 7.41, where the total gain G is set 2.6 by using a BEF and a DBEF. The BEF provides luminance gain in the normal direction and the DBEF provides energy efficiency gain through photon recycling. From the result, we find that the power consumption of white LEDs is 200 W when the luminous efficacy of white LEDs is 100 lm W^{-1}, the backlight efficiency is around 70% and the transmittance of the LC panel is 5%. When the luminous efficacy of the white LEDs is increased to 200 lm W^{-1} and the backlight efficiency reaches 90%, the required power of white LEDs will be only 78 W, and could be further reduced to around 40 W if the panel transmittance is 10%.

7.4 Optical design for other applications

White LEDs are applicable to general lighting and special lighting as well. In this section, we will introduce some special lighting systems using white LEDs, including dental light, marine light and tunnel light [48–50].

A medical lamp, such as dental light, should perform high-quality optical and color properties to meet the requested functions. In the past, tungsten halogens were applied

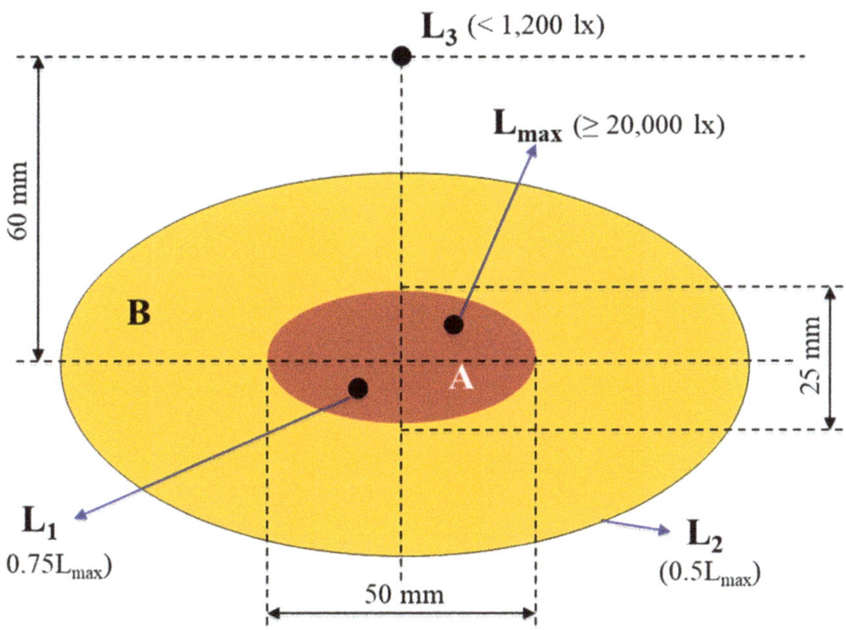

Figure 7.42. The illumination pattern requested in the regulation.

to dental light. But tungsten halogens are always operated at high temperatures. The intense infrared radiation could cause higher operation temperature. This could heat up a patient's mouth and cause discomfort. Another shortcoming is that the correlated color temperature (CCT) is fixed. Thus using a white LED to replace a tungsten halogen should be a good choice. The reference regulation for dental light is ISO 9608 [51], which indicates that the light pattern should cover an ellipse area when the projection distance is 70 cm from the dental light, as shown in figure 7.42. The light pattern should be in high uniformity. The minimum illuminance at the main spot should reach 20 000 lux and the biggest deviation of illuminance must be less than 25%. The illuminance in the outer area surrounding the main spot should be larger than half of the maximum illuminance. To avoid glare to the patient, the illuminance in some specific areas is limited to 1200 lux. Besides, the CCT is restricted to range from 3500 K to 6500 K, and the minimum CRI is 85 to ensure high quality of vision [52]. Also, the lamp must perform a minimum shadow pattern up to 12 mm on the target plane, and this requirement makes the lamp wider along the lateral direction.

To design such a lamp, one of the approaches is to build up a large reflector. The lateral size should be large enough to make the shadow effect acceptable. This design is suitable for tungsten halogen. When white LEDs are used in the lamp, one of the choices is to design an LED module with an optical projector for the requested pattern and to use two or four sets of the LED modules to form the dental light pattern. Owing to the ellipse pattern, an effective design is shown in figure 7.43, where one cylindrical lens and a double-convex lens form the projection optics.

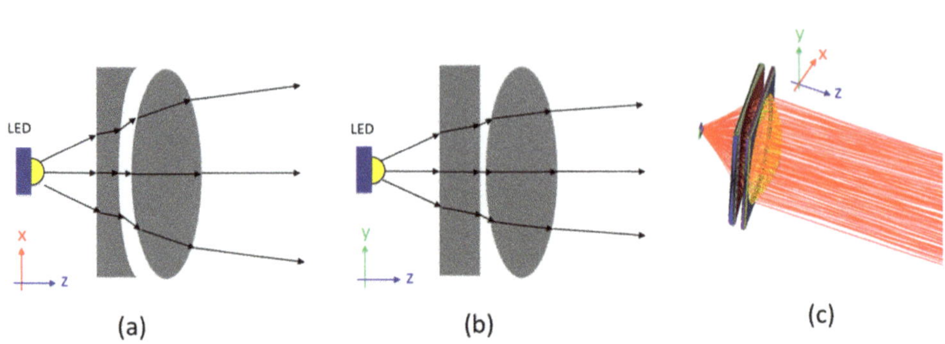

Figure 7.43. (a) The top view of the optical design for a dental light, (b) the side view, and (c) the ray fans.

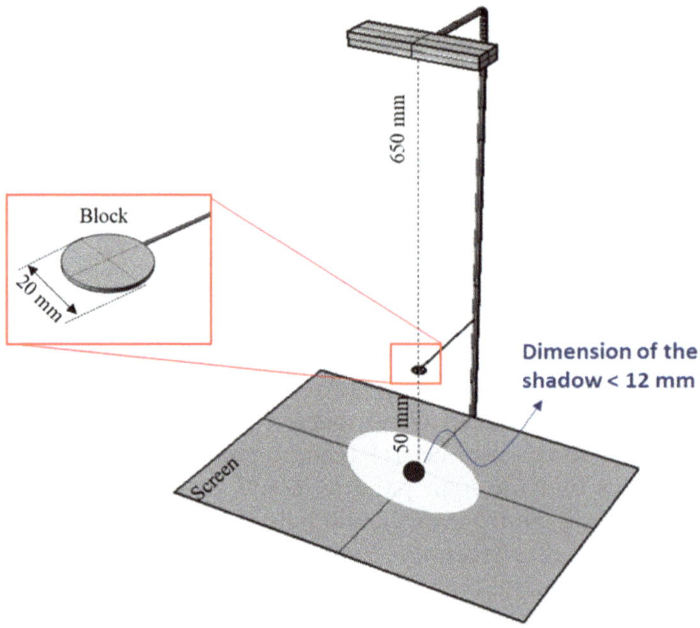

Figure 7.44. The testing configuration to check the shadow effect.

The double-convex lens is used to control the vertical width of the pattern. In the lateral direction, using a concave cylindrical lens to extend the lateral width.

To check the shadow effect, a structure is used, as shown in figure 7.44, where a round plate is located at a plane 50 mm above the target plane. Figure 7.45 shows a simulation for one LED module, two modules and four modules, where more modules spreading along the lateral direction will cause narrower shadow. Figure 7.46 shows the designed dental light containing four tilt optical modules and a prototype, and the designed and experimental illumination light pattern on the target plane. The maximum illuminance is 42 010 lux, which is around 85.7% of the designed value, when the injection power is around 9 W. Figure 7.47 shows the

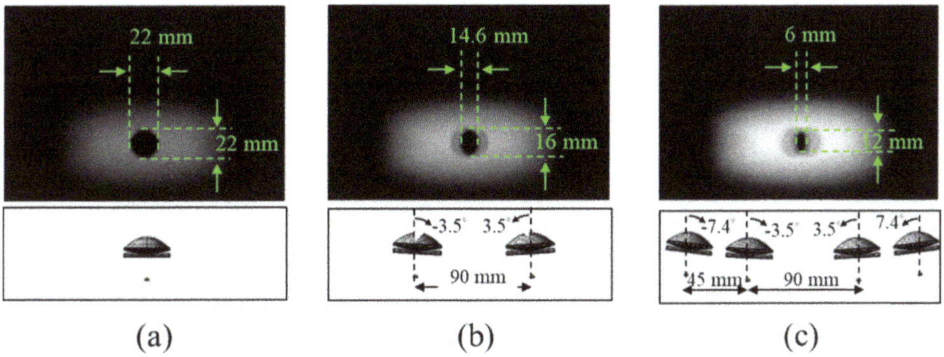

Figure 7.45. The shadow effects with (a) one, (b) two and (c) four optical modules.

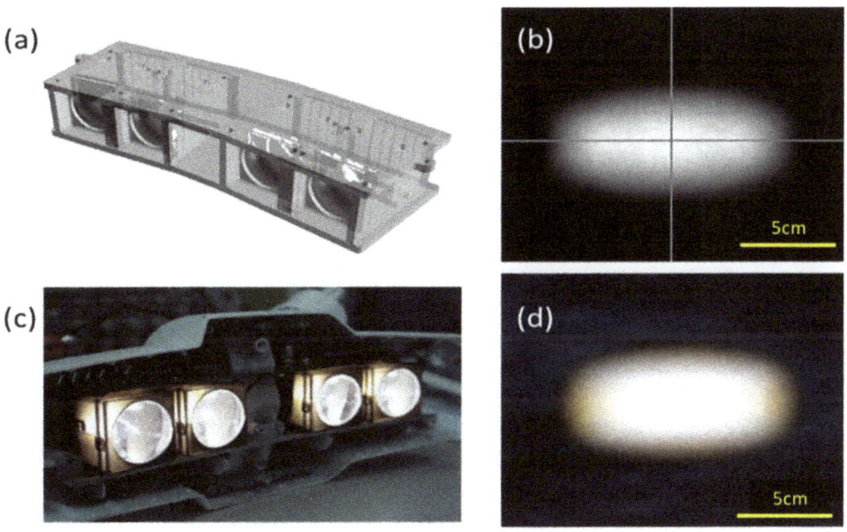

Figure 7.46. (a) The dental light structure, (b) the simulated light pattern, (c) the prototype of the dental light, and (d) the illumination pattern.

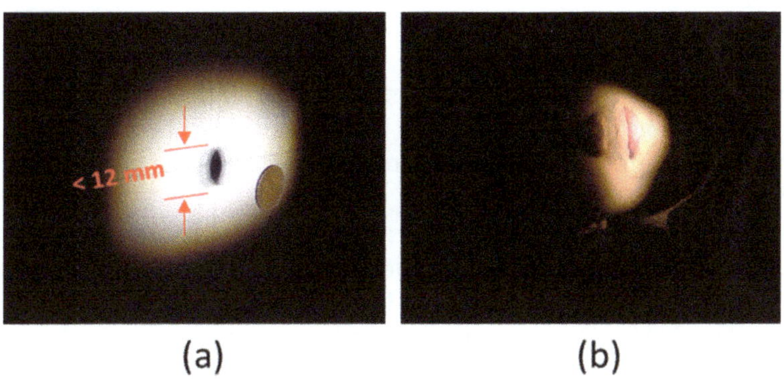

Figure 7.47. (a) The photo of the illumination pattern with a shadow, and (b) illumination on a patient's face.

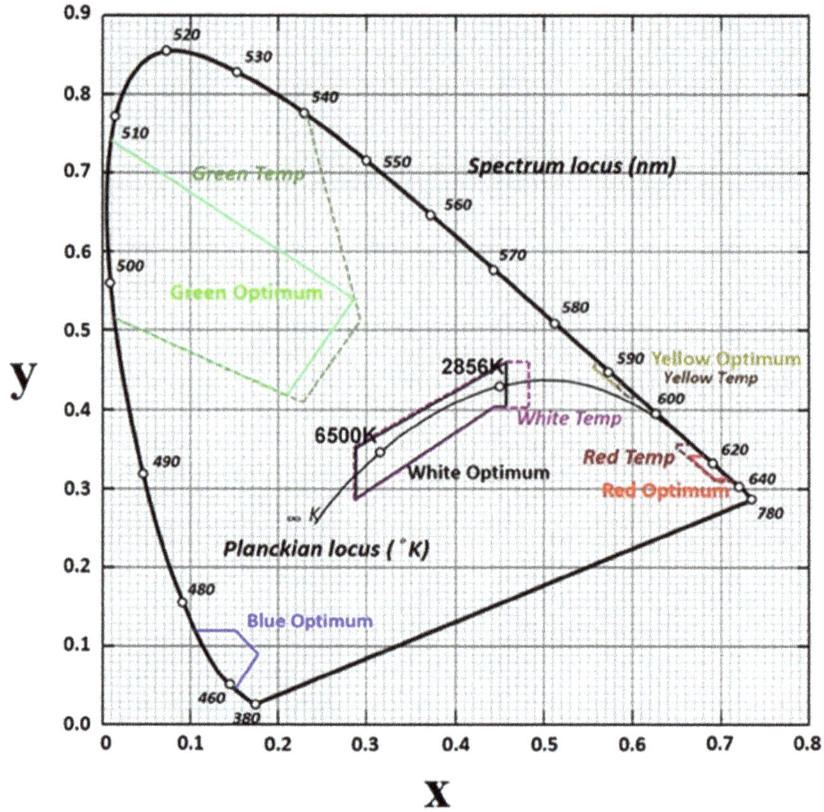

Figure 7.48. The color range definition of the IALA E-200–1 Marine Signal Lights marked in 1931 CIE chromaticity diagram [53].

experimental observation of the illumination pattern on a patient and the observed shadow in a measurement structure. The design principle of dental light is similar to other medical lights such as a surgical light, but a surgical light could require stricter dimensions in the shadow and color as well.

White LEDs as well as single-color LEDs are both suitable for marine application. For a marine signal light, the design could refer to the regulation for E-200–1 (Marine Signal Lights by the International Association of Lighthouse Authorities (IALA)) [53]. There are three main requirements, including divergent angle, projection range and chromaticity. In chromaticity, the range marked in 1931 CIE chromaticity diagram shown in figure 7.48 indicates the accepted color for the signal [54].

In the requirement of the projection range, the regulation considers the atmosphere transmission factor, meteorological visibility and the minimum visible luminous intensity with sky luminance. If the minimum projection range reaches 8 nautical miles (NM), for the unit atmosphere transmission factor of 0.74, the required luminous intensity is larger than 445 cd, and the full vertical angle must be at the range of 8° to 10°, and 360° in the horizontal direction. Such a requirement is

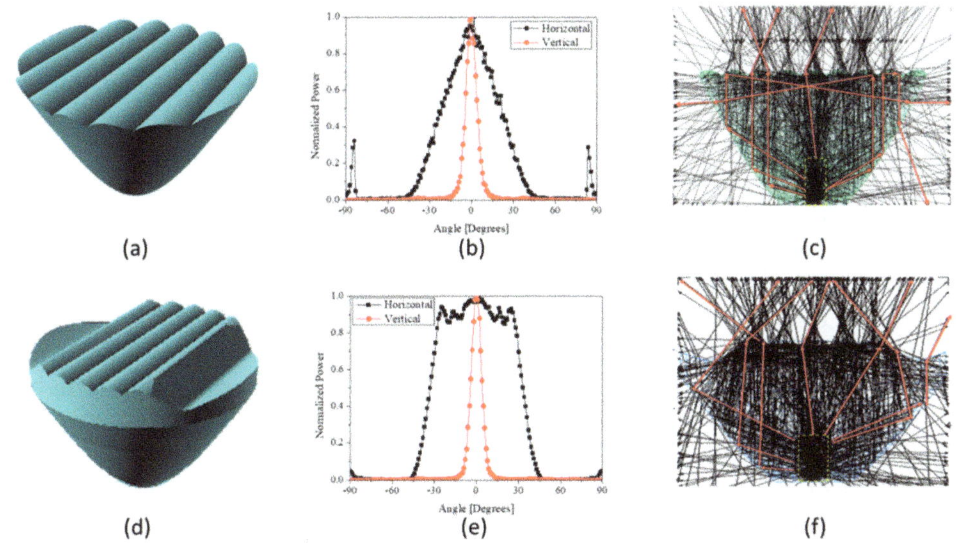

Figure 7.49. (a) The TIR lens with cylindrical lens array on the exit face, (b) the light distribution, (c) the ray fans; (d) the truncated TIR lens, (e) the light distribution, and (f) the ray fans. Reprinted with permission from [63].

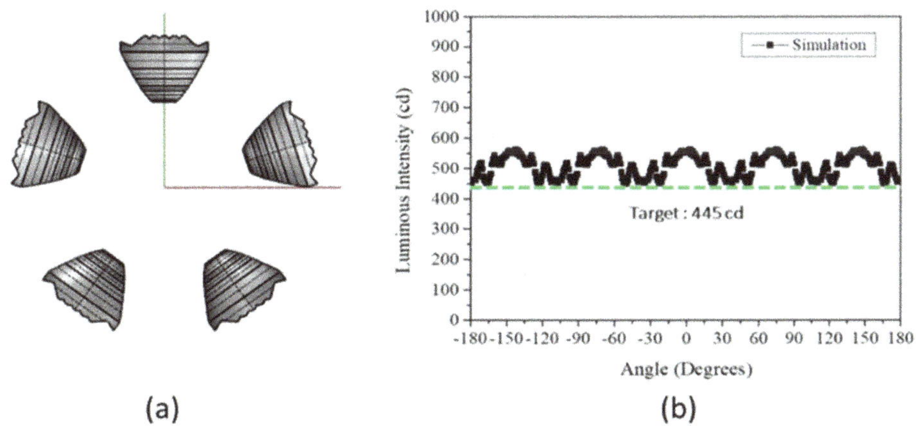

Figure 7.50. (a) The marine lamp with five LED modules, and (b) the simulated light distribution.

that angular distribution in the vertical direction is narrow and in the horizontal direction is wide. Along the vertical direction, it is suitable to use a TIR lens to collimate the light by the LEDs. Along the horizontal direction, it is applicable to introduce a cylindrical lens array across the exit face of the TIR lens, as shown in figure 7.49. Owing to more total internal reflection on the two horizontal edges, the two edges need to be truncated to make the incident light go forward. The edge-truncated TIR is shown with flat illumination within an angular range. The angular pattern along the horizontal direction is also important. Only flat illumination along the horizontal direction can make uniform illumination with multiple LED modules to make up a round illumination. Figure 7.50 shows the simulation of uniformity and intensity when five LED modules are applied, and the result meets the regulation demand.

A tunnel light can be regarded as one of the streetlights, but the design concept is somewhat different. In traditional tunnel lighting, a simple reflector is employed to collect the light downward onto the roadway. If the downward light pattern is too narrow, and the spacing of each of two tunnel lights is too long, there could be a dark zone between every two bright zones and a zebra fringe on the roadway. Besides, to avoid large brightness variation when driving in/out the tunnel, the number of the tunnel lights near the entrance and the exit should be much higher than in the other area. If a designer for the tunnel light only pays attention to roadway illuminance and energy saving, sometimes the tunnel light could induce glare when the light source is visible. To provide a comfortable tunnel lighting environment with low glare and high illumination uniformity, a proper optical design is necessary [55–58].

The optical design is aimed to provide uniform and bright illumination on the roadway, and to avoid glare to drivers. Then we can define a targeted area in the design, especially in regard to the glare effect, as shown in figure 7.51. The tunnel light is mounted on the ceiling 6 m above the roadway. Figure 7.51(a) shows that the front angle of the light radiation of the luminaire is set 65° along the driving direction and the back angle 22.5° along the reverse direction. The roadway is assumed to be a two-lane driveway with a sidewalk on each side. The width of each lane is 4 m, and each sidewalk is 1 m. According to this assumption, the target area is 10 m × 15.5 m, as shown in figure 7.51(b).

In the practical design, there are various ways to achieve the setting goal. Here we introduce a different light source, which is a white LED cluster (called WLEDC), shown in figure 7.52, where more than 100 pieces of blue LED are bonded in a high-conducting plate with covering yellow phosphor. The WLEDC is injected with hundreds of watt, so the heat dissipation is one of the important issues in handling such a light source.

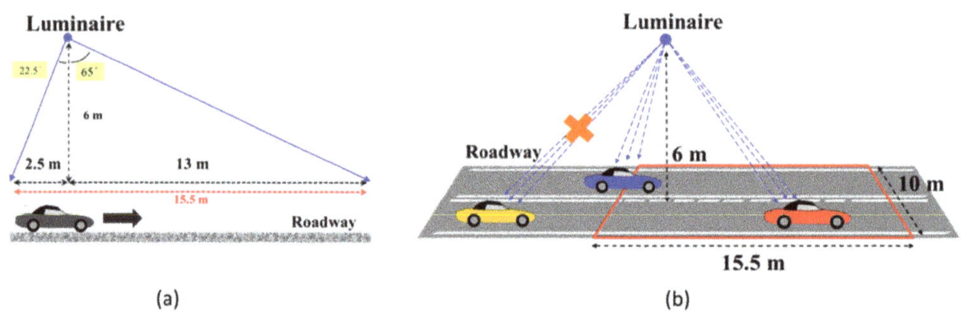

Figure 7.51. The illustration of (a) the control method of intensity distribution and (b) the target region for illumination.

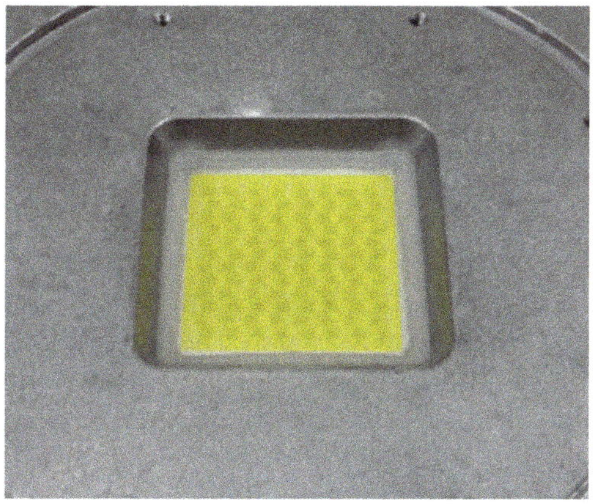

Figure 7.52. A photo of a white LED cluster attached on a metal board.

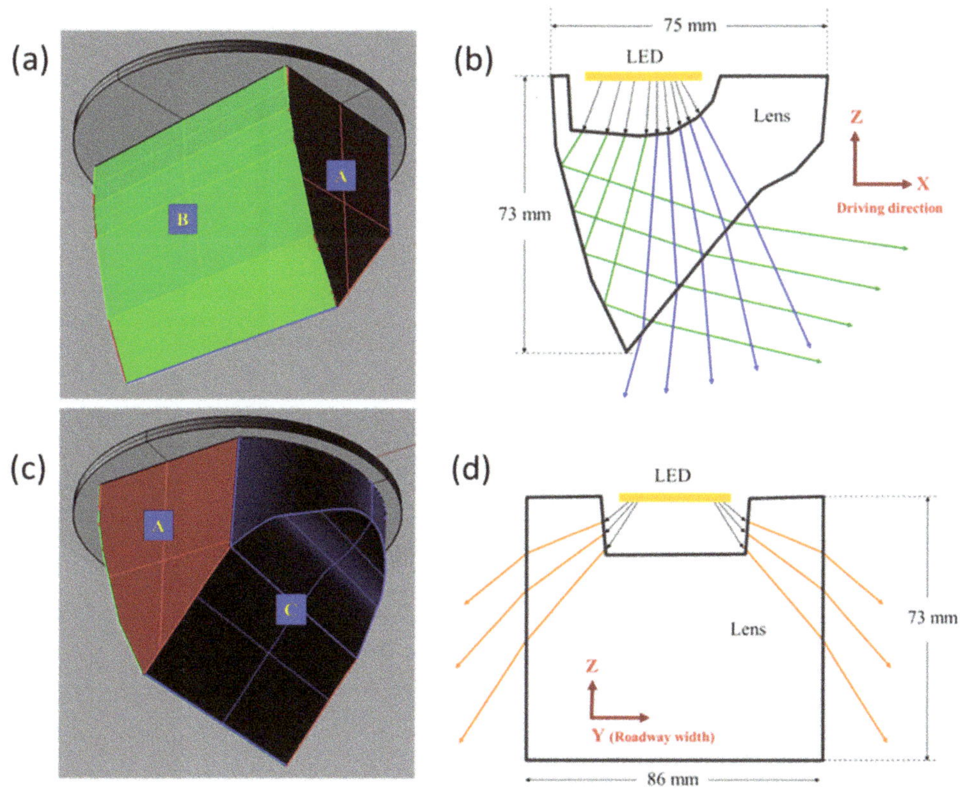

Figure 7.53. (a) The A and B faces of the freeform lens, (b) the light paths, (c) the A and C faces, and (d) the light paths.

To achieve low glare and high illuminance uniformity in the targeted area, an optical freeform lens is designed, as shown in figure 7.53. The design of the freeform lens is divided into two orientations, the *x*-axis direction and *y*-axis direction. The freeform lens comprises three parts, i.e., A, B and C faces. Owing to high operation temperature, the freeform lens is suggested to be made with glass, and thus the refractive index of the lens is set at 1.5. In the design, face A's are the side walls with less role in the illumination. Face B is to perform total internal reflection and to redirect the emitting light rays to the driving direction. The most important face is face C, which is to shape the illumination pattern downward onto the roadway from the direct rays by the LED. Also, face C needs to redirect the reflected light from total internal reflection to the front end of the target area. The simulated illumination pattern on the roadway is shown in figure 7.54. Figure 7.55 shows the angular distribution, where the designed projection light is directed along the drive direction, and therefore when a driver looks at the light source, it appears relatively dark, and no glare will be induced to the driver. The mockup sample and the measurement of the angular distribution are shown in figure 7.56. The simulation result fits the measurement well. The precise design and simulation is based on a precise model of light source. Since the LED is a big light source, the secondary

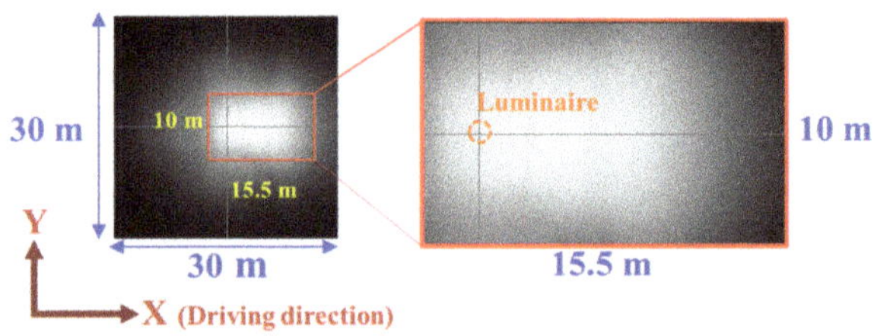

Figure 7.54. The illumination pattern on the ground.

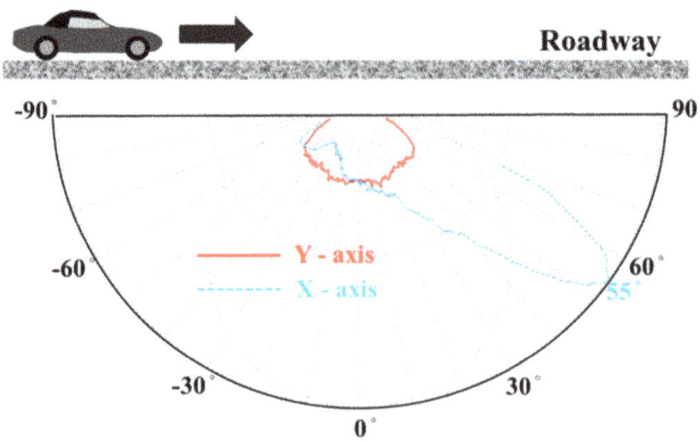

Figure 7.55. The light distribution along the driving direction.

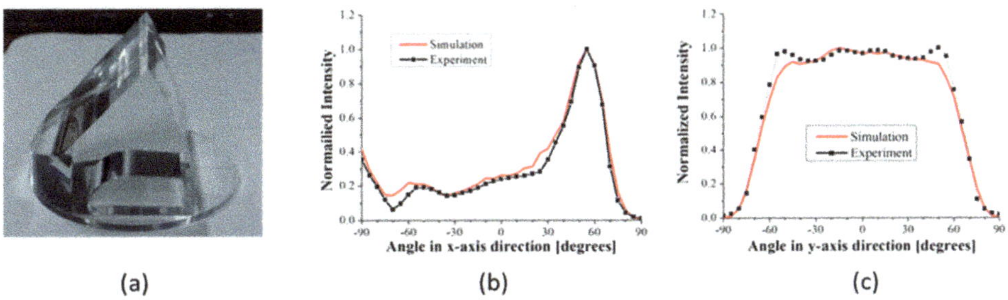

Figure 7.56. (a) The prototype of the freeform lens, (b) the angular distribution along the driving direction, and (c) along the other direction.

optics is located at the mid-field region of the light source. Here the mid-field model of a light source is important to make the design accurate.

References

[1] Sun C C, Jiang C J, Chen Y C and Yang T H 2014 Glare effect for three types of street lamps based on white LEDs *Opt. Rev.* **21** 215–9

[2] Luo Y, Feng Z, Han Y and Li H 2010 Design of compact and smooth free-form optical system with uniform illuminance for LED source *Opt. Express* **18** 9055–63

[3] Lo Y C, Huang K T, Lee X H and Sun C C 2012 Optical design of a butterfly lens for a street light based on a double-cluster LED *Microelectron. Reliab.* **52** 889–93

[4] Lee X H, Moreno I and Sun C C 2013 High-performance LED street lighting using microlens arrays *Opt. Express* **21** 10612–21

[5] Sun C C, Lee X H, Moreno I, Lee C H, Yu Y W, Yang T H and Chung T Y 2017 Design of LED street lighting adapted for free-form roads *IEEE Photonics J.* **9** 1–13

[6] Sun C C, Wu C S, Lin Y S, Lin Y J, Hsieh C Y, Lin S K, Yang T H and Yu Y W 2021 Review of optical design for vehicle forward lighting based on white LEDs *Opt. Eng.* **60** 091501

[7] UNECE Addenda to the 1958 Agreement (Regulations 101-20) http://unece.org/trans/main/wp29/wp29regs101-120.html

[8] Marelli Magneti Marelli Automotive Lighting Launches new LED Module Production in NAFTA http://magnetimarelli.com/press_room/news/magneti-marelli-automotive-lighting-launches-new-led-module-production-nafta

[9] EVS ISO 6742–1:2015 cycles–lighting and retro-reflective devices—part 1: lighting and light signaling devices https://evs.ee/products/iso-6742-1-2015

[10] Fujiyoshi S and Nakagawa T 2016 Lighting apparatus for vehicle and method of controlling a headlamp *Japan Patent 5* **941** 800B2

[11] Nakada Y and Ugajin Y 2015 Vehicular headlamp *US Patent* **9** 638B2

[12] Sun W S, Tien C L, Lo W C and Chu P Y 2015 Optical design of an LED motorcycle headlamp with compound reflectors and a toric lens *Appl. Opt.* **54** E102–8

[13] Chu S C, Chen P Y, Huang C Y and Chang K C 2020 Design of a high-efficiency LED low-beam headlamp using Oliker's compound ellipsoidal reflector *Appl. Opt.* **59** 4872–9

[14] Ishida H and Ken, S 2006 Vehicular headlamp and optical unit *US Patent* **7** 931B2

[15] Tatsukawa M, Ishida H, Sazuka K and Tokida T 2003 Vehicular headlamp *US Patent* **7** 453B2

[16] Wanninger M and Wilm A 2006 Lighting means having a predetermined emission characteristic, and a primary optics element for a lighting means *US Patent 20060083013A1*
[17] Tessnow T, Johnson R and Tucker M 2007 Application of LED light sources with light guide optics *SAE Tech. Pap.* **2007-1-1041**
[18] Tessnow T and Johnson R 2009 Led headlamp system *US Patent 20090034278A1*
[19] Sun C C, Wu C S, Hsieh C Y, Lee Y H, Lin S K, Lee T X, Yang T H and Yu Y W 2021 Single reflector design for integrated low/high beam meeting multiple regulations with light field management *Opt. Express* **29** 18865–75
[20] Sun C C, Lee T X, Ma S H, Lee Y L and Huang S M 2006 Precise optical modeling for LED lighting based on cross-correlation in mid-field region *Opt. Lett.* **31** 2193–5
[21] Chien W T, Sun C C and Moreno I 2007 Precise optical model of multi-chip white LEDs *Opt. Express* **15** 7572–7
[22] Sun C C, Chen C Y, He H Y, Chen C C, Chien W T, Lee T X and Yang T H 2008 Precise optical modeling for silicate-based white LEDs *Opt. Express* **16** 20060–6
[23] Cai J Y, Lo Y C, Feng S T and Sun C C 2014 Design of a highly head lamp with additional ground illumination *Light. Res. Technol.* **46** 747–53
[24] Lo Y C, Cai J Y, Chen C W and Sun C C 2012 A compact bike head lamp design based on a white LED operated at one watt *Opt. Laser Technol.* **44** 1172–5
[25] Lo Y C, Chen C C, Chou H Y, Yang K Y and Sun C C 2011 Design of a bike headlamp based on a power white-LED *Opt. Eng.* **50** 080503
[26] UNECE 2013 ECE324-R123 Uniform provisions concerning the approval of adaptive front-lighting systems (AFS) for motor vehicles
[27] Lee J H, Byeon J, Go D J and Park J R 2017 Automotive adaptive front lighting requiring only on/off modulation of multi-array LEDs *Curr. Opt. Photon.* **1** 207–13
[28] Yu H, Ro S J, Lee J H, Hwang C K and Go D J 2013 Smart headlamp optics design and analysis with multi-array LEDs *Korean J. Opt. Photon.* **24** 231–5
[29] Dubal P and Nanaware J D 2015 Design of adaptive headlights for automobiles *Int. J. Recent Innov. Trends Comput. Commun. (IJRITCC)* **3** 1599–603
[30] http://hella.com/)
[31] Liou Y C and Wang W L 2007 Lighting design of headlamp for adaptive front-lighting system *J. Chin. Inst. Eng.* **30** 411–22
[32] Ma R 2013 Automotive adaptive front-lighting system reference design (Texas Instruments)
[33] Unsolicited Opinions BMW adaptive LED headlights functionality https://youtube.com/watch?v=8Y3tkMD6Vew)
[34] Moreno I 2012 Image-like illumination with LED arrays: design *Opt. Lett.* **37** 839–41
[35] Basu C, Meinhardt-Wollweber M and Roth B 2013 Lighting with laser diodes *Adv. Opt. Technol.* **2** 313–21
[36] Wierer J J, Tsao J Y and Sizov D S 2013 Comparison between blue lasers and light-emitting diodes for future solid-state lighting *Laser Photonics Rev.* **7** 963–93
[37] Ulrich L 2013 Whiter brights with lasers *IEEE Spectr.* **50** 36–56
[38] Denault K A, Cantore M, Nakamura S, DenBaars S P and Seshadri R 2013 Efficient and stable laser-driven white lighting *AIP Adv.* **3** 072107
[39] Cantore M, Pfaff N, Farrell R M, Speck J S, Nakamura S and DenBaars S P 2016 High luminous flux from single crystal phosphor-converted laser-based white lighting system *Opt. Express* **24** A215–21

[40] Yang Y, Zhuang S and Kai B 2017 High brightness laser-driven white emitter for Etendue-limited applications *Appl. Opt.* **56** 8321–25
[41] Wu C, Liu Z, Yu Z, Peng X, Liu Z, Liu X, Yao X and Zhang Y 2020 Phosphor-converted laser-diode-based white lighting module with high luminous flux and color rendering index *Opt. Express* **28** 19085–96
[42] Sun C C, Lin S K, He M T, Wu C S, Yang T H and Yu Y W 2021 Enhancement of laser multiplexing efficiency for solid state lighting through photon recycling *IEEE Photonics J.* **13** 1–6
[43] Lee X H, Lin C C, Chang Y Y, Chen H X and Sun C C 2013 Power management of direct-view LED backlight for liquid crystal display *Opt. Laser Technol.* **46** 142–4
[44] Sun C C, Chien W T, Moreno I, Hsieh C T, Lin M C and Hsiao S L 2010 Calculating model of light transmission efficiency of diffusers attached to a lighting cavity *Opt. Express* **18** 6137–48
[45] Guselnikov N, Lazarev P, Paukshto M and Yeh P 2005 Translucent LCDs *J. Inf. Disp.* **13** 339–48
[46] Shieh H P D, Huang Y P, Lin F C, Chen H M P and Cheng Y K 2009 Eco-display an LCD-TV powered by a battery? *Dig. Tech. Pap.—SID Int. Symp.* **40** 228–31
[47] Lin F C, Huang Y P, Wei C M and Shieh H P D 2010 Color filter-less LCDs in achieving high contrast and low power consumption by stencil-field-sequential-color method *J. Disp. Technol.* **6** 98–106
[48] Sun C C, Ho H Y, Cai J Y, Chung S C, Chang Y Y and Yang T H 2015 Optical design of a high-performance and low-power dental light based on white LEDs *'Appl. Opt.* **54** E171–5
[49] Jen Y C, Lee X H, Lin S K, Sun C C, Wu C S, Yu Y W and Yang T H 2019 Design of an energy-efficient marine signal light based on white LEDs *OSA Contin.* **2** 2460–9
[50] Tsai M S, Lee X H, Lo Y C and Sun C C 2014 Optical design of tunnel lighting with white light-emitting diodes *Appl. Opt.* **53** H114–20
[51] ISO ISO 9680 Int. standard dentistry–operating lights (https://sis.se/api/document/preview/908781/)
[52] Hsieh C C and Li Y H 2014 The development of LED-based dental light using a multiplanar reflector design *Int. J. Photoenergy* **2014** 057079
[53] IALA IALA NAVGUIDE 2018 digital copy https://iala-aism.org/product/iala-navguide-2018-digital-copy/
[54] The Colour & Vision Research Laboratory Colour & Vision Research Laboratory and Database http://cvrl.org/)
[55] Lee X H, Moreno I and Sun C C 2013 High-performance LED street lighting using microlens arrays *Opt. Express* **21** 10612–21
[56] Painter K 1996 The influence of street lighting improvements on crime, fear, and pedestrian street use, after dark *Landsc. Urban Plan.* **35** 193–201
[57] Pachamanov A and Pachamanova D 2008 Optimization of the light distribution of luminaires for tunnel and street lighting *Opt. Eng.* **40** 47–65
[58] Haans A and de Kort Y A W 2012 Light distribution in dynamic street lighting: two experimental studies on its effects on perceived safety, prospect, concealment, and escape *J. Environ. Psychol.* **32** 342–52
[59] Lee C H 2015 The optical design of adaptive street lighting *MSc Thesis* National Central University

[60] Hu C M 2008 The design of automobile low beam by using white light LEDs with special package *MSc Thesis* National Central University
[61] Hsieh C Y 2021 Optical design and stray-light analysis for high-contrast low/high beam meeting various regulations *MSc Thesis* National Central University
[62] Chien K Y 2016 Optical design of adaptive automotive headlamp based on laser diode *MSc Thesis* National Central University
[63] Jen Y C 2016 The study of LED marine beacons *MSc Thesis* National Central University

www.ingramcontent.com/pod-product-compliance
Ingram Content Group UK Ltd.
Pitfield, Milton Keynes, MK11 3LW, UK
UKHW051855100926
8556IPUK00005B/629